Climate Change Apocalypse

By the same author:

Divine Weather

About the author:

Jack McGinnigle is a former Chief Forecaster of the Met Office, the United Kingdom government organisation for global meteorological science and services. His meteorological career spanned four decades, during which time he worked in many different locations in the UK, western Europe, North Africa and the eastern Mediterranean. His published scientific papers range across many aspects of meteorology, including original research into operational analysis methods. He also participated in the increase of production efficiency by developing technological strategies. To this end, he conceived the design of workstation-based visualisation and production systems for meteorologists; these are now used operationally.

Since leaving the Met Office he has retained an active interest in meteorology, lecturing on the weather and publishing further scientific work. As a Christian (he is a lay preacher in the Methodist Church), his first book combined scientific and biblical knowledge to examine how weather events, illustrations and concepts were used in the Bible as a remarkably effective communications media for people across the millennia. The book proposed several new interpretations of biblical texts and was widely discussed.

Climate Change Apocalypse

Jack McGinnigle

Highland

First published in 2010 by Highland Books Limited, 2 High Pines, Knoll Road, Godalming, Surrey, GU7 2EP

All quotations from the Bible are from the New International Version (1973) published by the International Bible Society

ISBN 978-1897913-85-7

Printed in the UK by CPI Mackays, Chatham ME5 8TD

Contents

This book is dedicated to
my grandchildren

List of Abbreviations

AD	Anno Domini
BC	Before Christ
CFCs	Chlorofluorocarbons
DegC	Degrees Centigrade or Celsius (temperature measurement)
DegF	Degrees Fahrenheit (temperature measurement)
DegN	Degrees North (position, latitude)
DegS	Degrees South (position, latitude)
EU	The European Union
GDP	Gross Domestic Production
in	Inches
IPCC	The Intergovernmental Panel on Climate Change
ITCZ	The 'Inter-tropical Convergence Zone'
GHP	Geothermal Heat Pump
kt	Nautical miles per hour
kph	Kilometres per hour
LPG	Liquid Petroleum Gas
mph	Miles per hour
Met	Meteorological
mm	Millimetres
m	Metres
m/s	Metres per second
sq.km	Square kilometres (area)
SSW	Sudden Stratospheric Warming
UK	The United Kingdom of Great Britain and Northern Ireland
UN	The United Nations
UNEP	The United Nations Environment Programme
USA	The United States of America
UV	Ultraviolet (radiation)
WMO	The World Meteorological Organisation

Introduction

This book has been written for the people of the Christian World, especially those who for many years have lived a comfortable and affluent lifestyle, with all their essential needs met. The 20th century saw many advances in science and technology which delivered significant improvement and efficiency to the world, particularly to those who lived in the more developed countries. Today, the rapidly growing technical advances of the 21st century continue the process. It is true that the negative effects of the world economic downturn of 2007 and the following years has affected everyone across the world. Even so, many Christian World citizens who live in affluence have been able to maintain their comfortable lives, especially when their lives are compared to those across the world who live a poor and precarious existence.

The Christian World comprises those areas of our planet where Christianity is the dominant or main religion. In geographical terms, it is very extensive and occupies large parts of the world. The major Christian World regions include the whole of North America (Canada and the United States of America), Central America and the Caribbean, all of South America, South Africa and much of Central Africa, the whole of Europe, much of Russia, Northwest China, Australia and New Zealand. In addition, a number of smaller Christian areas are to be found in India, the Philippines (both of these places have many millions of Christians) and other parts of Asia and the western Pacific. The adherents of Christianity comprise over 33% of the world population (Encyclopaedia Britannica 2005 survey), making Christianity the largest religion in the world.

Of course not everyone who lives in the Christian World is a follower of Christianity. Roman Catholic regions (mainly Central and South America along with some of the central and southern European countries) have the highest proportion of Christian followers within their total populations; in these areas, 90% or more are positively identified as Christians. Proportions in the protestant Christian areas (North America, the rest of Europe, Australia, New Zealand and Central/South Africa) tend to be somewhat lower, with over 75% of the populations designated Christians, while orthodox (Eastern) Christianity has a rather lower proportion still, perhaps nearer to 50%. These estimates were suggested by data published in the 'Adherents.com' website which assembles and publishes religious statistics obtained from sources all over the world.

For the purposes of this book, it is important to stress that it does not matter whether you the reader are a follower of Christianity or not, because all who are residents of the Christian World are required to conform to the laws and moral standards of whatever Christian country they occupy. In Christian countries, the laws and moral standards have been developed from the Christian religion; in other words, they have their basis in the Bible, the book of Christianity.

If you are a Christian, you will already be aware of this and are likely to be familiar with the biblical links which are presented throughout this book. On the other hand, if you have little or no contact with Christianity at present, it is still highly likely that you will recognise at least some of the biblical illustrations, either from earlier memory or from the fact that a great many biblical teachings and texts have been absorbed into everyday language and thought across the Christian World.

The book presents and explains the predictions of weather and climate change which flow from observations of a warming world. It shows how there are other mechanisms involved too. It seeks explanations, in particular whether human beings are responsible for the dangers that are predicted. It examines actions to solve the problem. While doing this, the book moves from science to economics, ethics and beyond. Throughout, the focus remains steadfastly upon the individual. The suggested thoughts and actions set out at the end of the book are not for governments or scientific authorities; they are for us, personally. Each one of us.

Every book which deals with a developing scientific reality will include facts and figures to support the arguments. There are many facts in this book. Facts specifically about weather, climate and global warming as well as energy usage realities in a changing and developing world. There are facts about the reasons why all these facts keep changing and what implications flow from these changes. There are also facts about the decisions and actions of humanity when faced with the dilemmas that have been presented to them particularly over the last two or three decades. Finally, because the focus is upon people of the Christian World, the facts extend into the morals and ethics of that particular world in the knowledge that the Bible has been a fundamental influence upon such matters.

Of figures there are deliberately far fewer and the reason for this is simple. This is not intended to be a scientific book giving detailed numerical data from the most recent studies. The figures attached to the whole complex arena that contains global warming, weather and climate change continue to be in a constant state of flux, frequently updated by later measurements, studies and interpretations; even historical data is subject to variation of opinion within the scientific community. Therefore numerical data (figures) are only presented in this book when they are judged essential to argument or understanding.

1

Where We Are Today

During the first decade of the 21st century, the marked and progressive strengthening of weather and climate change warnings has been very obvious. As the years have passed, the warnings have become stronger and more emphatic; as a result, national and international concern has grown across the world. Discussions and conferences have taken place at the highest political and scientific levels; international agreements have been negotiated and there are many plans for specific actions designed to stem the progression of the problems.

Today, it is notable that local and national politicians are eager to speak on the subject of 'climate change', adding it to their long list of problems that need to be tackled with urgency. Everyone is aware that climate change has become a frequent item in news broadcasts and is a popular topic of general conversation.

What do the warnings say?

The public of all countries in the world are familiar with the issue of predictive warnings by 'authorities'. In the case of scientific warnings, these authorities are often the research scientists themselves or the scientific organisation to which they belong; the warnings which they broadcast are based upon their own research work. The politicians enter the arena somewhat later, after the scientist's case has been accepted, to some degree at least.

In the case of climate change, this pattern has been followed and has taken quite some time to develop; starting from small beginnings several decades ago, the studies, theories and warnings have built up to become the powerful forces we are bombarded with now. Today, the warnings have acquired global authority; in many countries, perceived solutions have become law.

Since the latter part of the 20th century, the weather and climate change warnings have come from a truly scientific and prestigious source. They are issued and coordinated by the Intergovernmental Panel on Climate Change (the IPCC), which works under the auspices of the United Nations. The IPCC is a truly international body which includes many of the leading atmospheric scientists in the world. The warnings are also affirmed independently by many other leading scientific authorities, although there can be some differences in interpretation. It should also be noted that a minority of scientists disagree with the climate change predictions of the majority.

Over the years since the 1990s, the scientists linked to the IPCC have communicated warnings in a series of reports. From the start, the dangers of weather and climate change were linked to 'global warming', which is the term given to the observation that global temperatures on land and in the sea have been rising for most of the 20th century. During the last 20 years, the warnings have become stronger and progressively more refined.

The latest IPCC Warning:

Global warming, past, present and future

The latest warning was originated by the IPCC in 2007. The first part of the warning presents the global warming record to date, pointing out that during the previous 100 years, the 'Global Near-Surface Air Temperature' (the official title for temperatures recorded around 1.2 metres above the ground) has risen by 0.75

DegC, while the global sea temperature has increased 0.3 DegC on average over a depth of 1000ft or so.

This is followed by a prediction of future global warming. While earlier IPCC warnings always suggested that the continuation of the warming trend was 'likely' to some degree, the latest warning states that the evidence for continuing global warming is now 'unequivocal'.

It is also reported that these apparently modest temperature rises have already caused serious changes to occur across the world. Studies into polar sea ice in both the Arctic and the Antarctic have revealed considerable melting and break-up of the ice fields. Glaciers in many mountainous regions across the world have been observed to be subject to considerable melting and reduction in size.

The public will be familiar with this information since there have been many media reports on the subject in recent years, often expressed in the most dramatic terms. As a result of all this ice melting, world sea levels have risen and some very low level land areas of the world have already been permanently flooded. The sea level rise is predicted to continue and accelerate. Finally, the apparently modest rise in sea temperatures has already caused significant changes to plant and animal life on land and, particularly, in the waters of the seas.

Resulting predictions

The IPCC warning now turns to the prediction of change in weather and climate. These parts of the warning are expressed in terms of 'likelihood' or 'confidence' based on scientific predictions that have been formulated by trends in observations, scientific understandings and the interpretation of computer modelling experiments. Of these, computer modelling experiments have become increasingly important because the vast power and speed of today's computers now allow the 'models' to be highly sophisticated. Basically, these models are the equations which can represent atmospheric processes, and 'model runs' are often referred to as 'simulations'. In addition, the power of the computers permits huge quantities of data to be processed quickly.

On many occasions in the media and elsewhere, the warning has been given as one involving the possibility of 'climate change' alone but this book uses the expanded term 'weather and climate change' because this is a more accurate description of the range of the warning that the IPCC has issued. In fact it is now necessary to go further and detach weather from climate, because the remaining IPCC warnings actually address 'weather change' and 'climate change' separately. It is true that weather and climate are closely linked to each other, because the climate of an area or region is determined by the cumulation of weather events over a long period. So it is the cumulation of individual weather events that determine climate; weather events are the building blocks of climate.

– *Increased threat of extreme weather and related disasters*
  Turning next to the warning of 'weather change', the IPCC has reported over the years on its prediction that 'more extreme weather events' are likely to occur. Significantly, the predictions of the last two decades have reported a steady increase in the likelihood of this happening. The most recent warning suggests that 'more extreme weather events' are thought to be 'very likely'.
– *Abrupt or irreversible changes in climate cannot be excluded*
  The final warning addresses 'climate change'. Here, the prediction is setting out the likelihood of dramatic change, identified as 'catastrophic (sudden) climate change'. It has long been known that the climate of an area or region is affected by its proximity to the sea. The more conservative changes of temperature in sea water helps to impose a gentler climate on those land areas affected by the winds that blow from the seas. By contrast, it is a fact that land areas situated far from the sea will usually experience a harsher, more extreme climate.

  Importantly, the oceans of the world have organised sea currents within their waters. Where those sea currents are warm, for example the Gulf Stream, a notably mild climate will be imposed upon land areas affected by these currents. Catastrophic (sudden) climate change is linked to changes in these 'normal' sea currents. The disruption or weakening of such sea currents will impose a quite abrupt change of climate to the affected land areas. The

IPCC reports that such sea current changes are likely to occur to some degree but the computer modelling experiments currently suggest that a significant catastrophic change to major sea currents is 'unlikely' in the 21st century.

One matter that is of fundamental importance is the direct influence of humanity in weather and climate change. This has been hotly debated for decades with some scientists sure that we are all to blame for the problem while others refute any suggestion of blame completely. However recent research has now provided sufficient new evidence to convince the IPCC that humanity has a role in the problem. As a result, the IPCC reports have now pointed to humanity as a significant contributor to the current period of global warming. There are still howls of protest from some quarters!

Public reaction to the warnings

The full scientific reports issued by the IPCC to examine a range of warming scenarios that are calculated from various assumptions and input conditions. Thus there is a mass of data to consider and personal interpretation allows various groups of scientists to come to different conclusions about the precise nature of events and timescales. One effect of all this discussion and varied interpretation has been to 'muddy the waters' to some degree. This in turn has encouraged some people across the world to judge the whole matter as unpredictable, unimportant or not applicable to them.

Nevertheless the IPCC warnings have been accepted across most parts of the world, including the detrimental role of humanity. This is why the peoples of the world are being encouraged and directed to participate in strategies to mitigate the problem. That's right: whether we like it or not, we are getting the blame!

Most people acquire their information from the media with radio, television and newspapers probably the most dominant sources. On the subject of weather and climate change (and arguably everything else!), media sources tend to focus on the more dramatic facets of predictions; in consequence, frightening and depressing scenarios are often emphasised. This has certainly been the case in climate change reporting. When the warnings are paraphrased as media headlines, they are very likely to be expressed unequivocally, e.g. 'Global Warming Causes Climate Change'; this sort of headline is, at best, an inadequate simplification. Worse still is: 'Global Warming Has Changed our Climate'; this headline is not in accordance with the current IPCC warnings.

Furthermore, recent years have seen a marked expansion of the very human 'bandwagon' effect! Many in commercial or public life have been seen to become highly enthusiastic supporters of measures against global warming because this is in accordance with their business or political agenda. Obvious examples are manufacturers of alternative energy generation systems and pressure groups with 'green' and/or anti-pollution credentials. Equally, local or national authorities may use global warming arguments to support the imposition of new regulations, laws or taxes.

Nevertheless it is a fact that many countries have already taken mitigating action against global warming or, at least, have positive plans to do so. Some of these actions are regional and involve significant areas and populations. The strategy is to address what are seen as the root causes of the problem and to make every effort to 'save the world' from weather and climate change. It is of course absolutely necessary for large-scale actions of this type to be decided and coordinated by international and national administrations. However it is at the individual level that the actions will largely be taken. It is the individual who will be directed or, ideally, persuaded to take action. It is the individual who will reap the benefit if the effort is successful or suffer the consequences if it is not.

The trouble is that mitigating actions need to be applied globally. To be optimally effective they need to be taken and supported by every country in the world, especially by the larger, richer and more powerful states. Unfortunately, despite our 21st century knowledge of global interdependence, the nation states of the world are not always noted for their spirit of international cooperation. It seems that national or other interests often prevail.

This raises questions about the effectiveness of unilateral mitigating action, whether national or regional.

It is against this global background that the individual feels entitled to question why they are getting the blame and how effective and appropriate their personal actions are likely to be. In many countries, people are already aware that legislation will compel them to take certain energy saving actions; however, other actions they take may be the result of persuasion or individual choice. This is why everyone needs coordinated information about the fundamentals, issues and realities of all aspects of weather and climate change. With this information, people will be able to understand not only the current position but to evaluate the flow of new information that will undoubtedly appear in the future. Global warming, weather and climate change is not going to go away!

How does the Bible contribute to this topic?

The realities of the 'Christian World'

To a Christian believer, the Bible is a publication of great importance. It is the book of Christianity, the religion centred upon Jesus Christ. Its compilation of at least 66 individual books was written thousands of years ago by many authors, all believed to be inspired by God as they wrote. The purpose of the Bible is very diverse, setting out history, law and biographical detail. It is also full of essential teaching and instruction delivered directly or through story, poetry or parable. Notably, the Bible is full of love and morality. The Christian believes that God's will and purpose is revealed through these unique writings. Therefore the Bible is a huge influence on the life, decisions and actions of those who are active followers of Christianity.

However it is accepted that a significant number of people who live in the 'Christian World' have little or no direct contact with the Christian religion, although many may have had experience of Christianity in their earlier years. Also, a great many people have occasional contact with the Christian Church when they attend christening ceremonies for the children of family or friends; similar contacts occur with weddings and funerals.

It is a fact that many households in the Christian World possess one or more Bibles, even if they merely gather dust on a bookshelf somewhere. Certainly everyone in the Christian World is aware of the existence of the Bible and is likely to be familiar with some of its stories, e.g. the Creation, Noah's Ark, the Red Sea parting, Jesus walking on the water, feeding the 5000, etc.

The Bible is actually a mainstay of the Christian World

Many millions who live in the Christian World without meaningful contact with the Bible will no doubt be surprised to learn that it is actually a powerful force in their lives. This is because the Bible is the basis of the culture, morality and law of the whole Christian World, no matter where. The love and moral standards set out within the Bible have not only moulded Christian religious thought but influenced the structure of the legal, political and education systems that have been developed in the Christian World. It has also had a major influence on its languages, literature, art and music. Therefore, since every inhabitant of the Christian World lives under the laws and moral standards of their world, they are living with the laws and moral framework set out in the Bible.

The laws and morality of the Bible

The laws, teachings and advice are spread throughout the books of the Bible but are especially concentrated in the New Testament and particularly in the words and deeds of Jesus Christ. The following examples show a few of the teachings that have been a major influence in the Christian World.

The Ten Commandments[1] appear in a very early part of the Bible – Exodus is the second book of the Old Testament, immediately following the first book, Genesis. The Ten Commandments are also repeated, with some very slight variations, in the book of

---

1 Exodus 20:2-17

Deuteronomy, the fifth book of the Bible. The story of the delivery of the Ten Commandments is highly dramatic. Moses was instructed by God to climb to the top of Mount Sinai and there he was given the Ten Commandments inscribed on two stone tablets; the imagery of this story is well-known since it is one of the epic film maker's favourite biblical subjects! [1]

The Ten Commandments divide into two groups, the first dealing with a proper relationship with God and the second concerning proper relationships between people.

- *Commandments 1 to 4:* this group of four commandments teaches that there is to be only one God, that idols are not to be worshipped, that God's name is not to be used inappropriately and that the Sabbath is to be kept holy. These are important teachings for the Christian and, until relatively recently, they were largely obeyed throughout the Christian World. However social changes, especially latterly in the last century, have seen an erosion of these standards, especially affecting the teaching about the Sabbath. In most respects and in many places today, Sunday is no longer marked as a special day, different from the other six weekdays.
- *Commandments 5-10:* the second group of six commandments are easier to relate to the Christian World in general. These teach morality – what our attitude should be to others. It is significant that the first of these refers to the family, specifically that you should honour your mother and father. This is clearly linked with concepts of family discipline; indeed, at a deeper level, it is about love and respect. During recent decades in the Christian World, the actions of some children (and their parents) would seem to have drifted well away from this teaching, with regrettable and detrimental consequences for all.

The other five commandments are unequivocal; do not murder, commit adultery, steal, give false testimony (lie) or covet the possessions of others. These are fundamental and practical

---

1 Deuteronomy 5:6-21

rules for any civilised society. All the acts mentioned above are destructive and antisocial, invariably hurting individuals and society in general. This group of commandments specifies that you should act towards others with love, compassion and unselfishness.

The New Testament

The Sermon on the Mount is the longest speech that Jesus Christ makes in the Bible and it contains very important elements of teaching which follow on from earlier Old Testament material[1]:

- *The Beatitudes:*[2] the word 'beatitude' is taken from the Latin 'blessed' and the words of Jesus are a series of eight statements about specific types of people. Each statement begins: 'Blessed are.....' Five of the statements refer to people with problems: Those who have 'poverty of spirit', those who are in mourning, those who are meek or 'hunger for righteousness' and those who suffer persecution. Then there are three other statements about people with highly desirable qualities: Those who are merciful, those with purity of heart and those who are peacemakers. The Beatitudes teach and encourage a proper relationship of love and respect towards all people.
- *Love your neighbour as yourself:*[3] in the New Testament, Jesus spoke this teaching several times, referring to it as a 'Second Commandment' which followed the 'First and most important Commandment', which was an instruction to love God. Many people attribute the Second Commandment to Jesus himself but in fact he was quoting from Jewish scripture. The words of the Second Commandment first appear in Chapter 19 of the Old Testament book of Leviticus, one of the Jewish books of 'The Law'.

  However in other parts of the Gospels, Jesus expanded the Commandment further by linking it with the concept of praying

---

1 Matthew 5 to 7: all verses
2 Matthew 5:3-10
3 Leviticus 19:18, Matthew 22:36-39, Mark 12:29-31

for your enemies and 'doing good' to those who hate you.[1] Interestingly, in these expanded forms, the Second Commandment effectively replaces the Ten Commandments of Exodus; while most of the Ten Commandments instruct you on what NOT to do, the Second Commandment tells you what you SHOULD do. Therefore, if you obeyed the Second Commandment, you would not dream of doing any of the bad or negative things specified in the Ten Commandments – or any others like them!

Nevertheless, there is something of a problem here. The Second Commandment to 'love your neighbour as yourself' will only be effective if you love yourself – and some people do not love themselves at all. In fact some people dislike themselves, even hate themselves; in modern-day psychology, this is labelled 'low self esteem' and it is recognised as a serious medical problem. In these circumstances, the Second Commandment would certainly not fulfil its purpose. Jesus solved this problem by formulating his New Commandment.

- *The 'New Commandment' of Jesus:*[2] 'Love one another. As I have loved you, so you must love one another.' This is a very important teaching of Jesus. Here the Christian is directed to love <u>everyone</u> in the way that Jesus loved them. This takes away the problem of comparison with self. The Christian accepts that the self-sacrificial love of Jesus is infinitely intense and accepts that they must strive to replicate that love as best as they can.
- *The Lord's Prayer:*[3] everyone knows the Lord's Prayer as the very special prayer of the Christian faith. The words of the Prayer were given directly by Jesus Christ although subsequently additional words not spoken by him have been added to the beginning and end of the prayer. Today, this short prayer is spoken by Christians individually or often together in groups. It is regarded as a model for the essential elements of Christian prayer to God.

---

1 Matthew 5:43-48, Luke 6:27
2 John 13:34
3 Matthew 6:9-13, Luke 11:2-4

The influence of the Bible on the people of the Christian World

Law and life

The familiarity of these biblical injunctions show how the words and ethos of the Bible are firmly fixed in the structure of the Christian World and therefore in the lives of all those who live there. In practical terms, there are direct links with law; for instance, all law courts routinely require witnesses to swear to tell the truth while holding or touching the Bible (unless they practice another religion). Bibles are also to be found in hotel and other public rooms in all parts of the Christian World. These are usually provided by the Gideon Society who have placed over 30 million Bibles in this way. Many of these Bibles show clear signs of use.

The spiritual dimension

However, there is an individual spiritual dimension here as well. While it is true that many who live in the Christian World today claim no allegiance to God, it is not uncommon to see them turn to him readily, instinctively, at times of great stress or trouble in their lives. It is at these times they seek God's strength, help and comfort in the words of the Bible, through personal prayer or through others who are Christian believers. Whether they admit it or not, all people have been created as spiritual beings and their turning to God at times of greatest need is one clear proof of that. This is why the Bible is an important direct or indirect influence on the lives of all those who belong to the Christian World.

Prayer

The Bible is filled with prayers of all types. Prayers to God from his people, expressing their worship, adoration and love. Prayers for help and succour. Prayers for other people, known or unknown, seeking healing or restoration for them. In the Bible, there are a number of accounts telling how Jesus Christ taught people to pray. Significantly, he also taught that prayers will be answered but he stressed that a particular contribution was necessary from the praying individual. In Mark's Gospel he said: 'Therefore I tell you, whatever you ask for in prayer, believe that you have received it, and it will be yours.'[1] In this teaching Jesus is emphasising that faith is an important component of the process. When you pray for something, you must believe that you will receive what you ask for.

Prayer is an important means of communication that the Christian uses in his or her relationship with God. Of course prayer is available to everyone including those who have no apparent contact with God. In fact the word 'apparent' is very significant here, because although some people in the Christian World do not know God, he nevertheless knows them. Their prayers will be answered on the same basis as everyone else's.

Miracles

'That's an absolute miracle' we may say when something wonderful and unexpected happens. It is quite likely that we are not using the word 'miracle' in its pure sense, because the word is often used in everyday language. Nevertheless we have all seen miracles, real and genuine miracles. We have seen people saved when all hope was gone. We have seen wonderful events happen most unexpectedly. Sometimes, miracles have even happened to us. Most certainly, the answer to prayer is a miracle.

However, people have very different attitudes to miracles. Some people observe or experience these wonderful and unexpected events and wish to put them down to coincidence, chance, or luck (whatever that is!). Some will acknowledge miracles as the amazing events they are but do not wish to think about their origin. The Christian sees miracles all around and knows they are from God.

The Bible reports many miracles, all through its pages. Saving miracles of all types. Birth miracles. Healing miracles. Restoration to life miracles. Often delivered as a response to prayer. Christianity is full of miracles. Look around. They really are happening today.

1 Mark 11:24-25

The purpose of 'Climate Change Apocalypse?'

This book is not focussed upon any specific country or region because global warming is a whole world problem which affects all the people in the world. However there is a certain degree of focus upon the more developed countries because these tend to be the regions that use the most energy per head of population. It will be seen later in the book that the consumption of energy has a clear link to global warming. Therefore the information and advice in this book is directed firstly towards those who are more affluent, because these are the people who undoubtedly have the greatest power to influence the outcomes that are linked to global warming.

The book offers information on all weather and climate change issues and relates this knowledge to the choices and decisions that individuals will eventually need to make. In simple and compact terms, it describes the science of global weather change mechanisms and explores the possibilities which face the population of the world today and in forthcoming years. It looks at relevant historical factors, including the significance of the Bible and its teachings. It discusses the detail of the scientific warnings and the causes of foreseen problems. It presents the major global actions already in progress. It examines the links between the current and future problems and the actions of humanity. It maintains a personal focus so that people everywhere can explore their role and their personal responsibilities.

It is fortunate that many citizens of our 21st century world are remarkably well informed about the weather and its processes. This is largely to do with the information acquired from the comprehensive and frequent television weather broadcasts transmitted routinely by all countries in the world. Increasingly sophisticated and wide-ranging communications now ensure that such television broadcasts can be seen in all parts of the world, no matter how remote. In addition, the Internet is a rapidly increasing source of information and communication. Thus the weather expertise of people today is an excellent and appropriate platform from which to address the serious questions of weather and climate change presented in the following chapters.

2

This is Global Warming

As reported in the previous chapter, it has been known for some time that the world's land and sea temperatures are increasing. This increase has been recorded for much of the 20th century, since around 1920 or so. The warming continues into the 21st century. This had been noted with disquiet and concern, since at least two hundred years of records before 1920 have indicated largely stable temperatures in the world, with only relatively minor fluctuations in the record. This warming phenomenon of the last 90 years became known as 'global warming' in the scientific world and the term is now general knowledge.

Global warming is related to the concept of atmospheric balance. The atmosphere has various mechanisms to keep its elements in overall balance and it is only when these mechanisms are overcome that sustained changes occur. The 20th century warming record shows that the atmosphere was out of balance during most of that period because the world's natural temperature control mechanisms were unable to correct the effect of whatever input was (and is) responsible.

Why global warming happens

The imbalance which was the cause of global warming was studied scientifically during the latter half of the 20th century and eventually it was proposed by the majority of scientists that a build-up of atmospheric pollution of certain types was responsible for the air and sea temperature rises. Subsequently, humanity was identified as a significant contributor to that pollution.

The combined outcome of all the pollution effects involved proved very complex to determine; possible solutions and mitigation factors were even more so. Much work was done to define the precise sources of the most damaging pollution, to understand its full range of effects and to develop ways of reversing what are seen as very dangerous trends. This work continues today.

Atmospheric pollution through history

Atmospheric pollution lies at the heart of global warming and indeed it features in all other claims of weather and climate change. In the vicinity of large centres of human population worldwide, it has been obvious for many centuries that smoke emissions pollute the atmosphere. However, the human attitude to pollution was forged long before this.

Biblical pre-history: The story of humanity

Pollution is a harsh word that carries unpleasant meanings of contamination and defilement. It is a word often associated with the casual, careless and selfish activities of the human race. The Bible is thought to have been written over a period of 1500 to 2000 years, with the latest New Testament texts dated around AD100. Of course the early biblical texts are likely to be based on earlier ancient 'writings'. The Bible starts by describing remote prehistoric events; the earliest chapters of Genesis [1] refer to the creation of the world with no definitive timescale attached. Indeed, the timescale of 'The Creation'

1 Genesis 1-2: all verses.

ranges from the biblical six days[1] to a multitude of scientific propositions that stretch out to billions of years. The 'days' of the creation process of Genesis Chapter 1 are of course 'God Days' and so need not conform to the strictures of 24 hours!

After the biblical Creation accounts, the following chapters of Genesis [2] are directed squarely at the behaviour of Man. Here, the disobedience of mankind is described, starting with the poetic story of Adam, Eve, the Serpent and the Forbidden Fruit – this was when the 'apple' from the Tree of the Knowledge of Good and Evil was eaten. This was the first act of the 'Fall of Man' which brought punishment and separation from God.

The succeeding chapters of Genesis reported on a seriously deteriorating situation, with an increasing population routinely committing gross, selfish and detrimental acts in ways that were completely at variance with the perfection of God's creation. In time, these acts built up to great glut of 'wickedness'; as a result, the Bible tells us that God decided to destroy all this wicked life and start again.

Fortunately for us, one man is identified as 'righteous' and worthy of saving. So begins the very well-known story of the Flood and Noah's Ark.[3] First of all, the huge Ark is built to God's specifications, then it is filled with Noah's family and specimen pairs of all types of animals. When all is ready, the Flood commences and the Ark floats off with its precious cargo of life.

The story of Noah's Ark has a happy ending. Noah, his family and all the pairs of animals on board survive the Flood and, after many months, are able to disembark to dry land and begin to build new lives. At this time they are blessed by God who makes a very significant covenant (a binding agreement) with Noah; this stated that he would never again wreak destruction upon the life he has created.

---

1 Genesis 1:31
2 From Chapter 3
3 Genesis 6-9: all verses

The covenant was to be applied to all Noah's descendants and all the animals from the Ark. The sign and seal of this agreement was to be a brilliant bow of colour in the sky, the optical illusion we know as the rainbow. Sight of this miraculous feature would remind both God and all his people of the enduring and loving agreement, named the 'Rainbow Covenant'.

The fall of the people into 'great wickedness' [1] is a vivid illustration of the sort of casual, careless and selfish attitudes that are associated with human pollution. In this case, the word 'wickedness' is symbolic of a pollution of mind, body and spirit, the three fundamental parts which make up the totality of a human being. Sadly, the internal pollution (sinfulness) of mankind continued in the recreated human race and clearly continues to exist in our world today. However, the sin of human pollution is mercifully tempered by God's love and nurture, as promised by the Rainbow Covenant.

The biblical focus on human pollution

Smoke is a principal source of physical pollution and mankind has for millennia contributed to this problem. A word search of the Bible reveals that the word 'smoke' appears in English translations at least 45 times (the exact number depends on the translation). However none of the texts use 'smoke' as a specific description of human pollution. In the Old Testament (written originally in Hebrew), the Hebrew word for 'smoke' may also mean 'vapour' and there are a number of instances when the word is used to describe the concealment of the physical presence of God when he comes to Earth to communicate directly with his people. On these occasions, especially since there is often the drama of thunder and lightning present also, the writer's use of the Hebrew word is probably intended to be translated as 'cloud'.

Specific advice against pollution does appear, though. In Deuteronomy, within a passage devoted to personal and domestic hygiene, one of the laws given to Moses referred to strict

1 Genesis 6:5

anti-pollution measures that should be taken within the camp confines. It was specified that all human waste must be buried well away from the campsite. [1]

There is also a reference to a water pollution problem in another book of the Bible. In 2nd Kings,[2] the spring water of the city of Jericho had become polluted and was causing illness and death in the population. The cause of the pollution was unknown but clearly it was a very serious problem. The verses describe how the prophet Elisha cleansed the waters by using miraculous powers given to him by God.

Pollution worsens after biblical times (from AD100)

As the world population grew in numbers, the smoke of domestic, agricultural and industrial fires increased in line with the needs of larger populations. However, the population increase of the first millennium is judged to be rather modest; between AD1 and AD1000, the initial world population estimate of 200 million is thought to have increased to just 275 million (+37.5%) by the end of that millennium. This can be attributed to low infant survival rates, poor health and very much shorter life spans, as well as to the effects of violent conflict. From AD1000, the population increase accelerated and the next 700 years saw a doubling of the world population. During this time, there was an increasing concentration of people into urban groups.

It is certain that the populations of polluted areas would be aware of the detrimental effects of smoke and other pollution. However, coal and wood had to be burnt for domestic, agricultural and business activities and the people would have had no means of controlling the pollution effects. The situation worsened when heavy industry developed rapidly during the Industrial Revolution of the 18th and 19th Centuries; then, huge amounts of coal began to be used to generate energy. As a result, much larger volumes of smoke were emitted by workshops and

---

1 Deuteronomy 23:10-14
2 2 Kings 2:19-22

factories, greatly augmenting the steadily increasing contributions from domestic chimneys.

This pollution was not only an unpleasant inconvenience but also a serious medical hazard. Down the ages, the lives of many have been affected detrimentally by constant smoke inhalation and ingestion. Many have been poisoned in this way or suffered from respiratory or other diseases that eventually caused their death, often at an early age. Again, there were few anti-pollution solutions on offer – only taller factory chimneys to encourage the smoke to be carried further away – to affect someone else!

Smoke and carbon

Smoke is an effluent of burning and the constituents of each sample of smoke are determined by the materials being burned. This means that every sample of smoke will have a different mix of chemicals within it, each with its own potential hazard. Because carbon is a fundamental material in our world, this chemical element is found almost invariably in smoke; indeed it is usually the sooty carbon particles that make smoke visible.

The human race has been producing smoke for many millennia and so releasing into the atmosphere carbon and other chemicals in both solid and gaseous forms. Within all that smoke, there is one gas always present; this is carbon dioxide, formed when one atom of carbon combines with two atoms of oxygen. Carbon dioxide is formed during most energy-producing processes. This includes not only all the commercial and industrial processes that spring easily to mind but to the very existence of animal life itself. Our bodily energy-producing processes involve the exhalation of some carbon dioxide with every breath; 4-5% of what we breathe out is carbon dioxide.

Carbon dioxide: Friend or foe?

In recent decades, carbon dioxide gas has been identified as one of the major causes of global warming and so has been labelled one of the villains of the situation we are now in. All over the world, people have been advised that they must reduce their personal output of carbon dioxide. Fortunately, this does not mean that you should breathe less! It means that we should all be mindful of global warming and its effect on weather and climate and so take action to reduce activities that are linked to carbon dioxide output.

Strangely, carbon dioxide is actually fundamental to life, since living organisms are built on the conversion of carbon dioxide into carbon based organic compounds. The maintenance of our planet depends upon the 'carbon cycle' which recycles the element in various ways and at various rates. One demonstration of the cycle can be seen in the interaction of animal and plant life.

Animal life requires oxygen as biological fuel and emits carbon dioxide. Plants absorb carbon dioxide, remove the carbon from it and emit oxygen. This process is called plant photosynthesis and it is responsible for most of the oxygen in the atmosphere. Plants (everything from algae to the largest trees) are thus an absolute essential for animal life on Earth. So carbon dioxide, although in volume only a minor component of the Earth's atmosphere (currently around 0.038% by volume), is very much an essential aspect of life.

The net effect of the complex carbon cycle is to maintain a constant amount of carbon dioxide in the world system. This is done by attempting to balance negative and positive carbon dioxide flow rates (like the animal/plant relationship described above). Inevitably, there are flow-rate imbalances and these are dealt with by the creation and maintenance of world carbon reservoirs. For instance, the world's oceans are large and very important carbon reservoirs which absorb a large proportion of atmospheric carbon dioxide into their waters. On land, forests are the most important stores of carbon, the largest of these stores being found in the vast forests of Russia and the rain forests surrounding the length of the River Amazon in South America.

The trees of life

Mankind has always regarded trees as very important living organisms. Like so many other plant forms, trees share an important role as a human food source in the form of a wide range of fruit; in some cases, leaves may be eaten, too. Early Man certainly used trees for shelter and security, since living above ground in a tree provided some protection against ground-based predators.

However the most important product of trees has long been wood and mankind has made use of this immensely strong and versatile material for an incredibly wide range of purposes. Today, we still do.

Just a few centuries ago, the scientists of the day discovered a new dimension in the importance of trees when they realised the essential link between trees and the life cycle of most animals. As already stated, all animals emit a small but significant proportion of carbon dioxide in their breath. Plant life, including trees, absorb this carbon dioxide, process it through photosynthesis, retain the carbon and emit oxygen into the atmosphere, replenishing the oxygen animals use. Nowadays, the vast rainforests of the world are recognised as major players in this process; this is why large-scale deforestation is now regarded with such great concern.

For most people, trees also have some sort of spiritual dimension. Here is a large complex organism sprouting from the soil of the earth and thrusting upwards towards the sky, sometimes for hundreds of feet. We see most trees as things of beauty, strength and elegance. To remind us of this beauty, we paint them and photograph them to adorn our walls. We admire their shape and structure; the thick and powerful central trunks, the powerful spreading branches that taper to delicate ends. We wonder at the biological mechanisms within that raise sap and nutrients to great heights and at the mystery of the fertility that is represented by the seeded fruits they bear.

We rejoice that the wood of trees is so universally useful for mankind's purposes. We are thankful that it can be used for so many processes, everything from the most delicate and intricate carvings to huge and powerful structures in a myriad of shapes and forms. If at this moment you are reading the printed word, this will be printed on paper, a product which is usually manufactured from wood.

The singular trees of the Bible

The Bible also reflects humanity's spiritual and practical connection with trees. The word 'tree' or 'trees' appears in the Bible over 300 times, almost always connected to positive imagery or implication. Tree imagery is often linked with peace and contentment, with love and nurture. The book of Micah gives a good example: 'Every man will sit under his own vine and under his own fig tree and no one will make them afraid.'[1]

Trees first appear in the Creation story in the first verse that mentions plants and vegetation.[2] There is also an early reference to the beauty and practicality of trees: [3] 'And the Lord God made all kinds of trees grow out of the ground - trees that were pleasing to the eye and good for food.' These particular trees were pictured in the Garden of Eden, the idyllic place created as the home for Adam.

More significantly, this early part of the Bible goes on to describe two particular trees which play a very important part in the biblical story of the Fall: 'In the middle of the garden were the Tree of Life and the Tree of the Knowledge of Good and Evil.'[4] In that very well-known story, God forbad Adam to eat from the Tree of the Knowledge of Good and Evil, saying:[5] '... for when you eat of it you will surely die'. Eve and then Adam subsequently disobeyed God and ate the fruit. They then suffered the consequences of banishment from the Garden of Eden and separation from the nurture of God.

The other tree mentioned in the description of the Garden of Eden, the Tree of Life, does not appear in the story at this point. Subsequently, however, it is revealed as the fundamental reason for the harsh act of banishment. The Bible text says:[6] 'And the Lord God said, "The man has now become like one of us, knowing good and evil. He must not be allowed to reach out his hand and take also from the Tree of Life and eat, and live forever".' So the banishment took place and life became very much harder

---

1 Micah 4:4
2 Genesis 1:11
3 Genesis 2:8-9
4 Ibid.
5 Genesis 2:17
6 Genesis 3:22

for Adam and Eve, although their stewardship responsibilities for the world were unchanged.

A literal interpretation of these events presents a picture of extreme human weakness and wilful disobedience, followed by the swift and harsh retribution of God; however, this interpretation is suggested to be simplistic. If the complete story of God and mankind is followed throughout the Bible, the account of the Fall of Man is more likely to be a powerful spiritual image which links to the other end of the Bible in the book of Revelation, the last book in the New Testament. In its latter parts, the book of Revelation sets out predictions for the future and a promise of total redemption for mankind, a redemption which will return Man to the perfection of God's protection.

The quotation from Genesis 3:21-22 (above) reveals that the fruit of the Tree of Life imparts immortality. Having eaten from the Tree of the Knowledge of Good and Evil, Adam and Eve had introduced sinfulness into their life. Eating fruit from the Tree of Life would then make them and the human race which follows from them eternally sinful. In the story, God intervenes to prevent this from happening, since he did not wish his perfect creation to acquire immortality until they were without sin.

Subsequently, it is the life, death and resurrection of Jesus Christ that makes eternal life possible again. In the book of Revelation, the promise of eternal life is revealed. In one passage,[1] the resurrected Christ speaks to the Church at Ephesus, urging them to repent and promising: 'I will give the right to eat from the Tree of Life, which is in the paradise of God.' Thus the promise of this essential concept of Christianity spans and pervades the whole Bible.

The 'greenhouse effect'

Carbon dioxide also has a very important role in the world's weather – it acts as a 'greenhouse gas'. The way greenhouse gases affect atmospheric temperature is known as the 'greenhouse

1 Revelation 2:7

effect'. Today, greenhouse gases are usually regarded negatively since they are considered to be responsible for the dangers of global warming.

The greenhouse effect alters the conditions on the surface of the Earth in just the same way as a greenhouse alters the conditions inside its structure. The garden greenhouse achieves a heated climate by means of its panes of glass. The sun's radiation flows unrestricted through the clear glass heating the floor, plants and benches within. The air in the greenhouse is then heated by contact (conduction) and spread around by air currents (convection) to make the whole internal atmosphere hot. The hot air is prevented from leaving by the panes of glass. All greenhouses, from the smallest garden unit to the largest commercial structure, work in this way.

The atmosphere produces the same effect without panes of glass. Here, greenhouse gases are carried upwards from near the Earth's surface to accumulate invisibly in the high atmosphere. The sun's radiation is able to pass virtually unrestricted through this cloud of greenhouse gas because solar radiation has a very short wave-length. (Radiation wave length depends on the temperature of the radiating body; the hotter the object, the shorter its wave-length.) So in a cloudless situation, the energy of the sun's radiation is practically undiminished when it arrives at the surface of the Earth. There, it heats up the Earth's surfaces which, in turn, heat the air above them by conduction. The heat energy is then spread upwards by convection currents and air turbulence.

Meanwhile, the heated ground surface also radiates heat energy but, because its temperature is very much lower than that of the sun, its radiation wave-length is much longer. The greenhouse gas cloud in the high atmosphere is not transparent to this longer-wave radiation. When the Earth's radiation reaches the greenhouse gas layer, some of the radiation is absorbed by the gas cloud, some is reflected back to Earth and some escapes to Space though the gas cloud. Thus the lower-level atmosphere where we all live is kept much warmer that it would otherwise be if there were no greenhouse gas cloud. The more dense the greenhouse gas cloud layer becomes, the greater the greenhouse effect and the hotter the lower-level atmosphere becomes.

In fact the greenhouse effect is yet another essential for our lives. If the effect did not exist, it is certain that the Earth would be considerably colder. 'No greenhouse effect' calculations suggest temperatures at least 35 and possibly as much as 50 Centigrade degrees colder. This means that planet Earth without the greenhouse effect would be a deeply frozen world with no possibility of life as we know it. In fact, it is important to realise that it is the greenhouse effect that saves us from being in a permanent Ice Age.

Greenhouse gases: Not just carbon dioxide

- *Water vapour:* although carbon dioxide has already been identified as a major greenhouse gas, there is another atmospheric gas that actually makes a much greater contribution to the greenhouse effect. This gas is water in its invisible gaseous form. Water vapour is a small but variable component of the Earth's lower atmosphere, present in every sample of air. In extremely hot and humid conditions, it can be as much as 5% of the volume of air but normally it is in very much lower concentrations, often less than 1% at the surface of the Earth and very much less in the upper atmosphere.

  The presence of water vapour in the atmosphere is an essential for most of the weather we experience. If there were no water vapour in the atmosphere, the Earth's weather would be confined to variable windiness and local obscuration by dry particles (sand, dust, soot, etc.) caught up in the wind. There would be no clouds, fog, mist or water-based precipitation of any kind.

  As a greenhouse gas, it is calculated that the small concentration of water vapour actually contributes up to 70% of the total greenhouse effect but, because it is a naturally-occurring greenhouse gas whose effect has not changed (or been changed) over millennia, it does not usually feature in discussions about global warming.

  After water vapour, carbon dioxide is next in importance as a greenhouse gas. The effect of carbon dioxide outstrips the other greenhouse gases by a considerable margin, since its contribution is around 26% of the total greenhouse effect. The other significant greenhouse gases are methane, nitrous oxide and various forms of chlorofluorocarbons (CFCs), all present in the atmosphere in very much smaller volumes. On the other hand, the heat absorbing potential of these other greenhouse gases is very much greater than carbon dioxide; the methane absorption value is 21 times greater than carbon dioxide and nitrous oxide 270 times greater. All together, these other greenhouse gases contribute only 4-5% of the total greenhouse effect, about one-sixth of that contributed by carbon dioxide.

- *Methane:* colourless, odourless and abundant in nature, methane is the main constituent of the natural gas used by us for energy-creating purposes. It is also used directly in many manufacturing processes and can be processed into other useful chemicals. It is produced naturally when plant matter decomposes underwater, hence its traditional name 'marsh gas'. Significant quantities of methane are also released from solid landfill sites and from livestock farming activities. Coal contains methane which can become a hazard in coal mines when it is mixed with air, forming an explosive gas known in mining circles as 'firedamp'.
- *Nitrous oxide:* one of several oxides of nitrogen, it was used for a time as an anaesthetic for brief surgical operations. Unfortunately it is poisonous and more prolonged inhalation causes death. A more modern use of nitrous oxide has been as a propellant gas in aerosol spray cans. It is also emitted by a number of agricultural and industrial processes.
- *Fluorocarbons:* this group of gases do not occur in nature. They have been manufactured in a number of forms, the most widely used of which were several formulations of chlorofluorocarbons, referred to generically as CFCs. These gases contain carbon, chlorine, fluorine and hydrogen in various compositions. They only came into use during the 1930s when they began to be used as aerosol spray can propellants.

In subsequent decades the use of these CFCs became extensive, being useful in refrigeration and many manufacturing processes. In many ways they seemed to be ideal products because they were neither toxic nor flammable and were extremely versatile in use. Then it was discovered they were extremely powerful greenhouse gases. By this time, related forms of other equally powerful greenhouse gases had been produced, such as hydrofluorocarbons, perfluorocarbons and sulphur hexafluoride.

As greenhouse gases, these manufactured substances played only a very small part in the greenhouse effect because their atmospheric concentration was so very small. However it later became apparent that they posed a very serious threat to the environment in a completely different way. At very high (stratospheric) levels, the presence of tiny concentrations of these gases acted to destroy atmospheric ozone. The implications of this effect are very serious and this will be discussed fully in a later section.

Out of balance: why greenhouse gas concentrations have risen over time

Graphs charting the concentrations of carbon dioxide, methane and nitrous oxide in the atmosphere all illustrate a similar increasing pattern in the recent past. The current concentration of carbon dioxide in the atmosphere is around 0.038% by volume; compared to oxygen for instance (21% of air is oxygen), the carbon dioxide concentration is tiny. Methane concentration is almost 200 times less than carbon dioxide and nitrous oxide much less still. Nevertheless these very small concentrations are significant.

Until the 18th century, when the Industrial Revolution began, it seems that greenhouse gases in the atmosphere had been present in largely constant quantities for an extended period of time. In the case of carbon dioxide, it has been possible to measure trapped samples of the gas in ice samples more than 1000 years old; these measurements have shown that there were only small fluctuations in the gas concentrations until the mid 18th century. During this long period of time, the concentration value was measured at around 0.027%.

The situation then changed to indicate a slow but steady increase during the 200 years between 1750 and 1950. From 1950, the graphs steepened markedly and this steeper increase has been maintained to the present day. The build-up of the other naturally-occurring greenhouse gases, methane and nitrous oxide, show the same sort of increase patterns.

Of course the build-up of CFCs presents a completely different picture since these manufactured compounds were not present in the atmosphere before 1930. From zero concentration the graph rises very steeply during the next 60 years. However, as far as lower level warming is concerned, the tiny concentrations of CFCs greatly limit their effect when compared to the effect of carbon dioxide. This is just as well, because CFCs are extremely powerful greenhouse gases.

So why did greenhouse gases increase like this? The carbon dioxide balance mechanisms have already been described earlier in the chapter; the balance is maintained by positive and negative flow rates plus the adjustments of carbon reservoirs. From the time of the Industrial Revolution, the situation changed quite markedly. Suddenly there was much greater energy consumption and a consequent increase in carbon dioxide production.

At the same time, the accelerated felling of forests for wood fuel directly decreased the absorption of carbon dioxide by plant life. In addition, the reduction of the forests physically decreased the size of the land-based carbon stores. All these actions combined to destroy the balance potential and the outcome was a steady increase in carbon dioxide concentrations in the atmosphere.

From 1950, the post World War II days saw a further marked jump in global energy consumption as rising prosperity and the advances of medicine and technology gave the populations of the developed world (at least) the confidence and the means to consume energy as never before. In these days, it is a matter of record that little attention was paid to energy conservation in either domestic or industrial matters. Energy was cheap and plentiful. The result was of course a consequent increase in carbon dioxide production as more natural fuels (coal, gas, wood) were consumed.

As this occurred, just like the earlier Industrial Revolution situation, the same rapidly developing technology that consumed energy and produced carbon dioxide made the process of deforestation easier and quicker. It became possible to penetrate remote wooded areas like the huge rainforests and to take large-scale machinery there. The reduction of vast forests reduced carbon dioxide plant absorption even more and the land carbon reservoirs shrank.

So it is clear the process that started in the 18th century has continued and intensified. Today, global levels of carbon dioxide are increasing by around 0.5% per year. The process continues and may well accelerate. In addition, recent reports also express concern about the condition of the rain forests themselves. There are worrying reports that individual trees are showing signs of health problems; this too would reduce the overall efficiency of the forests.

3

Why Blame Us?

For several decades, the IPCC has kept us informed about global warming. They have advised us about the consequences, specifically what is happening to sea and glacier ice and what may be happening to weather and climate. They have described the vast forces that are at work all over the world, from the depths of the sea to the heights of the stratosphere. Then, they have told us it's our fault! Why?

Didn't global warming happen before?

This is often the first question people ask when accused of being responsible for the current round of global warming. The answer to this question is undoubtedly: 'Yes.'

Global warming has happened before, several times, in fact. Paleoclimatologists, who study ancient climate, have mapped out for us how the Earth has cycled through ice ages and very hot periods, within a record that stretches back beyond 2,000 million years. Their assessments are derived from geological study, ancient sea ice samples and (for more recent times) from dendroclimatology, where trees up to 5000 years old reveal significant information about the local climate in the growth ring patterns clearly visible in sections of the tree trunks.

The last very hot period is said to have occurred during the Cretatious Period, which is thought to have ended around 65 million years ago. Of course mankind in its present form or number was not in existence then, so cannot possibly be blamed for this period of hot conditions!

Does the Bible have anything to contribute to this discussion?

A word search for 'global warming' in any current Bible translation is certain to offer no results. Searching for 'climate change', 'carbon dioxide' or 'greenhouse gas' will have the same effect. Many people think that there cannot possibly be any 'science' in the Bible – and, yes, that's another word that does not appear! However the fact is, people are quite wrong about the lack of science in the Bible. There is a lot of weather in the Bible and some of it is remarkably scientific.

The people who wrote the Bible showed not only an awareness of weather and climate but, at times, an understanding of their workings. For instance, many biblical characters are identified as farmers and fishermen and these activities have always been significantly weather-sensitive. For such people, knowledge of local weather can be a matter of success or failure in their business; for some, it can be a matter of life and death at times, too. The Bible presents many illustrations of weather knowledge. There is even quite sophisticated weather science to be found.[1]

So although the term 'global warming' is not to be found in the Bible, the concept is certainly there. When God imposed punishment or destruction upon his people, he frequently used the natural forces of the world; quite often, these were extreme elements of weather. Storms, floods, tempests and thunderbolts were all popular! For instance, in the very first biblical destruction

1 The author has explored this subject in detail in his book 'Divine Weather' published by Highland Books ISBN 1-897913-61-3

event, the Flood, huge quantities of rainwater were used to flood the Earth deeply.[1]

If this rainfall was to be generated in ways that meteorology currently understands, incredibly large amounts of heat would be required; only this could hope to produce the energy needed to drive such a rainfall development. The link between heat energy and rainfall is already demonstrated on our Earth; the most violent rainfall is produced consistently in the Tropics where Earth temperatures are very high and there are plentiful supplies of moisture. So there is a remote theoretical possibility that the Flood could have been initiated by extremely severe global warming.

The story of the destruction of the cities of Sodom and Gomorrah provide another piece of biblical evidence which shows God's use of extreme heat. The evils taking place in these cities are well-known and God decided that destruction was appropriate. After saving Lot's family, most Bible texts refer to 'fire' raining down. However the Hebrew word written originally is defined as 'fire, literally or figuratively' so there is again just the possibility of a link with global warming, if a scientific explanation is being sought.[2]

Yet another example is to be found in the Old Testament book of Isaiah. In a section entitled 'The Apocalypse of Isaiah', the final downfall of the Earth is announced. Here, the terminology of destruction includes the prediction that the Earth will 'dry up and wither' and its inhabitants will be 'burned up.' This suggests strongly that great heat and drought will occur, so linking again with concepts of global warming[3].This theme of destruction by great heat is repeated later in the book. [4] There is also a reference to extreme heat affecting the Earth in the New Testament. In this

---

1 Genesis 7:11-12
2 Genesis 19:24
3 Isaiah 24:4-7
4 Isaiah 51:6

passage, it is stated there will be fire which will lay the earth bare and destroy everything.[1]

The Revelation of St John is an unique book which concludes the New Testament and thus is the last book of the Bible. The content of this book includes letters, drama, prophesy and revelation. The word 'revelation' is the English translation of the Greek word *apocalypse* which refers to the unveiling of secrets about Heaven and Earth. These are revealed through the text written by the apostle John and through the reported words and actions of Jesus Christ and God's angels.

No other book in the New Testament contains apocalyptic text although some words of Jesus in Matthew's gospel may be interpreted as a prediction of St John's revelation. Certainly, Matthew gives unsettling details of 'the End Time'.[2] Some Old Testament books have elements of apocalypse within them, for instance, Daniel and Ezekiel.

The apocalyptic writings of Revelation present a series of awesome images involving fire, heat and total destruction. Fire was hurled upon the Earth, accompanied by thunderstorms and hail and the Earth was burnt up.[3] Finally, there is the most direct link with a global warming concept when the sun's power was greatly increased and delivered very intense heat to the Earth, to 'sear' and 'scorch' everything, both living and inanimate.[4]

So it seems that God used the concept of severe global warming in past millennia and could do so again. Remembering that the latter parts of the book of Revelation are predictions about the future, particularly about the 'End of This Age', some argue that the global warming we now see could be the start of that process.

---

1 2 Peter 3:10
2 Matthew 24:29-41
3 Revelation 8:5-7, also 16:18
4 Revelation 16:8-9

Our Fault? Where global warming is different this time

Today's populations have been identified by the IPCC as significant contributors to the current round of global warming. The scientific case is set out below.

Measurement and collection of data

The starting point of any scientific investigation is reliable data. It was not until the advent of scientific measurement in the 18th century that air temperatures near the Earth's surface could be recorded reliably and average values calculated with reasonable accuracy. Even so, in these early days of measurement, observations were relatively restricted in number and suffered from instrument and recording inaccuracies. In addition, there were considerable problems in collecting the data; for instance the records from ocean-going sailing ships would generally not be obtained until the vessels returned to port, often months later.

In the 19th century, the invention of radio greatly improved the data collection situation and the accuracy of scientific instruments was also much improved. The 20th century brought even more accuracy, much faster communications networks and, latterly, remote and automatic measuring equipment and numerical manipulation of the data by electronic computers. Since the 1970s, orbiting weather satellites have joined in to provide millions of temperature measurements over land and sea, adding to the values that are still provided by land and ship-based human observers all over the world.[1]

The global temperature record

The global near- surface average temperature for any particular year is arrived at by taking average daily temperatures from a huge range of locations on land and at sea; these are then averaged again to produce a single value. Today's global average temperature is

1 Appendix B: Weather satellites see p. 218

calculated to be around 15 DegC. This is a meaningless number when considered in isolation but it becomes significant when compared with values from earlier years; then it becomes possible to see how the global temperature has changed over an extended period.

For the data reliability reasons explained above, most graphs start around the mid-19th century. It is normal for scientists to calculate temperature 'anomalies' – that is, values that show how far the temperature for a particular year varies from a past average; so the anomalies can conveniently be expressed as positive or negative values that are equivalent to 'warmer' or 'colder'.

Everyone is thoroughly familiar with the variability of weather experienced during any year and it is equally obvious that each year is different from previous ones. This is true even of regions where there are only relatively small variations in weather during the seasons; in such places, the population is attuned to small variations in weather factors. For instance, when the weather is extremely hot, a difference of one degree can be noticeable; 42 DegC can feel much hotter than 41 DegC!

Of course the variations can apply to a number of other weather factors such as rainfall, sunshine, windiness and of course, to levels of temperature. The annual temperature calculated for an area or region can allow the year to be judged 'cold', 'average' or 'warm' in accordance with the temperature anomaly for that year. A statistical device to further smooth out the effect of annual fluctuations is to calculate five-year rolling averages and compare these. These five-year mean values of global temperature are useful for showing longer term trends and so are very suitable for global warming considerations.

The five-year mean values of global temperature calculated for the period from 1850 to the current day present a record which can be analysed into four distinctive time periods. The first period comprises the seventy years between 1850 and 1920, during which time the temperature record was in a fairly stable situation, showing mainly slight variations around the norm.

There then followed a period of 30 years (1920-1950) when there was a clear and sustained rise in the global average temperatures. By contrast, the following fifteen years (1950-1965) presented steady or slightly falling average temperature values. Thereafter, 1965 to 2009 showed a clear record of increasingly positive values; during this time, the global average temperature was rising at an increasing rate. These records were presented in the IPCC 2001 report.

Temperature and carbon dioxide graphs compared

Since the increase in carbon dioxide concentration is considered to be a major factor in global warming, the temperature and carbon dioxide graphs could be expected to show a similar shape.[1] However, when the two graphs were compared it was immediately clear that they were dissimilar in one important respect. While both graphs show fairly stable values until 1920, then a significant increase in values during the period 1920-2009, the temperature rise (described in the paragraph above) checked and even decreased a little during the period 1950-1965, while carbon dioxide concentrations continued to rise during that period. So although the average global temperature graph showed a generally similar rising pattern to the carbon dioxide concentrations graph, there was a period in the middle of the record (1950-1965) when the records were dissimilar to each other.

This dissimilarity caused a good deal of controversy in the latter parts of the 20th century. However a resolution of this problem has since been proposed and it is discussed in the following sections. Today, the IPCC and the majority of scientists working with this problem are satisfied that their warnings of rising sea levels, weather and climate change are real and justified. However there have always been detractors and, although now much fewer in number, they continue to disagree with the conclusions of the majority.

---

1 Such graphs are widely available and updated on the Internet.

Global warming today: The scientific propositions refined but also disputed

The build-up of global warming evidence

When global warming was first proposed three or four decades ago, it was suggested that the build-up of greenhouse gases was responsible, in particular the increase of carbon dioxide. The increase of greenhouse gas concentrations was subsequently attributed (in part) to the activities of humanity worldwide – but more especially to those in the 'developed world' where resource consumption was much greater per head of population. This was just a theory at the time, mainly because there were few past records of the amount of carbon dioxide in the atmosphere; the main source of evidence came from the analyses of air trapped in ice core samples taken from polar regions or mountain glaciers. Since then, carbon dioxide and other pollutants have been measured routinely.

Of course it has taken time for science to assemble the temperature record and to seek the precise causes for the variations it shows. However many scientists have long been convinced of the link between greenhouse gases and global warming. For the last two decades at least, they have warned that the ultimate outcome of global warming will be to change the weather and climate of the world in a significant way – at least in some areas. For this reason, they have recommended that action should be taken to reduce greenhouse gas emissions. In the 20th century however, there was a significant amount of disagreement from a number of sources - scientific, political and individual. Additionally, from many ordinary citizens of the world, the reaction was disinterest and apathy. 'The world is capable of looking after itself' was a popular response when any intervention was proposed.

Significant confirmation in the 21st century

Since global warming was first recognised as a problem, the number of scientific dissenters has declined, especially in more recent years. This is undoubtedly due to the build up of evidence which has convinced the scientists associated with the IPCC studies to strengthen and refine their warnings. The basis of the warnings given over the years has been the observation of the changes in global temperatures, especially with regard to the melting of ice volumes, and the ability to model weather and climate changes using complex computer simulations.

Climate model computer simulations have always been dependent upon reliable data and the accuracy of the mathematical models that could be applied to that data. In recent decades, the explosive increase in the power of computers has been obvious to all, since everyone can now hold awesome computing power even in small hand-held devices. Half a century ago, such computing power, even if available, would certainly have occupied a very large building and required a large cooling plant to prevent the huge banks of electrical components overheating. The increase in computing power has contributed to a greater understanding of atmospheric processes and this in turn has allowed much greater sophistication to be built into the computer simulations of atmospheric processes.

Temperature is an important datum in meteorology, traditionally obtained by reading a thermometer. 'Surface' temperatures are usually measured in the shade (usually within a ventilated white painted box) set around 1.2m above the ground. A huge number of individual observations have been made all over the world during the 24 hours of every day and night. Such observations continue to be made today.

Over the centuries, many temperature records have been assembled into what we now call 'databases'. Eventually it was possible to combine these to become world temperature databases and several have been assembled and maintained over the decades. In the latter part of the 20th century, great efforts were made to improve the quantity and quality of past data. Today, three world near-surface temperature databases are maintained, one in the UK and two in the USA. Although each database uses different methods of construction, all show the record of a warming world over the 20th century and beyond.

In 2005 the latest database was used as the initial input for a series of highly sophisticated climate models experiments. The purpose of these experiments was to see how accurately the models could predict today's reality from the starting point of the historical records. If this could be done successfully, it would enable a better understanding of the global temperature rise which occurred in the 20th century. In addition, it would indicate whether the greenhouse gas pollution of humanity was a factor in the global warming process.

There is also an important future element to these experiments. If the simulations managed to predict today's reality from a historical start-point, the whole process could then be repeated with today's data as the start-point and the future predictions of global warming could then be regarded with increased confidence. Importantly, the result would then suggest what action could be taken to provide mitigation or solution. The process and implications of these experiments are very important and are described in the next section.

The computer simulations

It has long been known that 'natural processes' contribute to global warming. These include the effects of large-scale natural pollution (volcanoes, extensive forest fires caused by heat or lightning) and extra-planetary effects like changes in the sun's radiation. None of these natural events are connected in any way to the activities of mankind. To this 'natural' global warming is added the effects of 'human' global warming, caused by fossil fuel burning and the other pollutants for which we are responsible.

The experiments used the most sophisticated climate computer models then available and the 'new' historical database in an attempt to establish the mechanism that produced the global warming patterns that have been recorded over the last 100 years. So that the warming from both natural and human pollution could be studied, simulations were run with three different atmospheric input conditions, that is, including natural pollution only, then human pollution only and finally with both pollution sources combined.

The first simulation (including only natural pollution) suggested a modest warming effect until 1950, a slight decrease in the 1950s and early 1960s, then returning to a slight warming record thereafter. The second simulation (using only human pollution effects) showed modest warming to 1950, a hesitation from 1950 to 1960 then a steeply increasing rise to the present day. The third simulation included both natural and human pollution effects. The result was a global average temperature record that was dramatically similar to reality. The result of these experiments were hailed a significant success and have been generally accepted in IPCC circles, with the data published in recent reports.

Understanding the results

The generally stable global temperature record from 1850 to 1920 reflected the fact that the natural carbon dioxide processing systems (e.g. the plant/animal oxygen/CO2 interaction) plus the balancing effect of the available carbon stores on land and sea were able to keep atmospheric carbon dioxide values largely constant; although there were fluctuations from year to year, there was no significant warming or cooling trend.

The modest warming trend between 1920 and 1950 showed that the increase of carbon dioxide and other greenhouse gases was sufficient to overcome the atmospheric balance and the result was an increase in the global average temperature value. The increase of carbon dioxide comprised the combination of natural effects (volcanoes, changes in solar radiation, etc.) plus the greenhouse gas production of a world population which was numerically less than one-third of today's figure; the 2 billion world population of 1920 rose by 50% to reach 3 billion in the 1950s.

There was a decrease in natural global warming effects in the 1950s but the human contribution to greenhouse gases was sufficient to maintain the global average temperature around existing levels for 15 years or so, although there were some positive and negative fluctuations in the individual values.

From 1965 onwards, 'natural' global warming was generally low so whatever warming took place from this year was produced largely by human input. The rapid increase in human greenhouse gas emissions during this period is explained by the combination of a quickly expanding world population and greater affluence in many parts of the world. This led progressively to much more fuel resources being consumed with the inevitable consequence of increased greenhouse gas pollution.

Since then, the world population has continued to increase rapidly, more than doubling since the middle of the 20th century to reach today's figure of around 6.8 billion. Increasing populations, continuing affluence in many areas of the world and the expansion of technology of all kinds meant that energy was being consumed as never before. The amount of greenhouses gases produced by the peoples of the world continued to increase, global warming was boosted and the global average temperature values soared.

In the opinion of most scientists today, the evidence presented by these experiments has now established a very strong (some say irrefutable) link between the global warming pattern and the production of greenhouse gases by the activities of the world's population. This is why the consensus of scientific opinion worldwide in 2010, through the communications of the IPCC and others, has proposed positively that humanity now contributes in a significant way to the current situation of global warming. The contribution is shown by increasing concentrations of greenhouse gases, largely the result of burning fossil fuels for a range of energy creation purposes.

Of course it will come as no surprise to hear that there are those who continue to disagree!

Scientific dissent

In all human life, there are always dissenters. It seems to be part of the human condition. In science, dissent is common. With every scientific theory published there will always be those who disagree. They dispute the theories proposed, the evidence gathered, the interpretation of such evidence, the observations made and the conclusions drawn. They often propose alternative theories. Sometimes these alternative theories are based on different interpretations of the same evidence, which are then followed to reach different and often opposing conclusions; at other times their alternative theories are based on different evidence and research.

Where the science is involved with prediction, the dissent is likely to be more extensive, because prediction always involves assumption and interpretation. This is certainly the case for global warming, weather and climate change. Since the problem was identified there have been many publications of dissent. Today, because of the progress of the mainstream scientific work that has taken place within the IPCC umbrella, dissenters are fewer in number than they were ten or twenty years ago – but they are still vociferous!

The following paragraphs give details of some of the major arguments of the scientific dissenters which have been published over the decades of global warming investigation:

– *The rebalancing argument:* Two or three decades ago, a popular area of dissent in the global warming argument concerned the shape of the global average temperature record. When the 1950s hesitation in the rising temperature record was plotted, some scientists pointed to this and claimed this was evidence of the atmosphere rebalancing itself. Therefore, they proposed that global warming could be ignored as a danger.

  However, the rapid and sustained temperature rise of the following decades acts powerfully against the rebalancing proposal. Furthermore the hesitation in the global temperature rise can now be explained by separating the effects of natural and man-made greenhouse gas production. This was achieved by recent computer simulations, described earlier in this chapter.

– *The involvement of cosmic rays:* some dissenters have claimed for decades that worldwide temperatures are controlled by variations in solar cosmic rays. Their proposal states that the variation of cosmic ray levels is directly linked to the amount of cloud cover in the world and this in turn affects the global temperature. So their claim is that the global warming now in progress is not linked with the increase in greenhouse gases at all and so is completely outside the control of humanity. This particular theory has been explored for several decades and finds few supporters.

- *The human contribution is insignificant:* it has been proposed by some dissenting scientists that the human contribution to the greenhouse effect is extremely small and therefore insignificant. They have stated that the accumulation of greenhouse gases includes a very large contribution from the effect of atmospheric water vapour; furthermore, that the presence of cloud in the atmosphere adds to the warming of the lower atmosphere, too. Their calculations suggested that mankind's direct contribution (by burning fossil fuels, etc) was very small - possibly less than 1% of the total effect. Therefore they claimed that the global warming already recorded was very largely the result of the events identified as 'natural'; such events are unconnected with the activities of mankind. Therefore they propose that attempts to alter global warming are pointless.

  This proposition has been rejected by the majority of scientists who have always pointed out that the increasingly steep global warming record after 1965 accords with the rapid human population build-up and the associated increase in post-war affluence. This explanation of the warming record was supported powerfully by the recent climate computer simulation experiments described earlier in this chapter. Using historical data, these simulations were able to replicate the reality of the temperature structure during the last 90 years only when both natural and man-made effects were included.

- *Temperature measurement disparities:* the procedures for measuring temperatures near the surface of the Earth have already been mentioned. However, meteorology also requires to know the vertical temperature structure through the atmosphere so special temperature measuring instruments are carried routinely through the atmosphere, lifted by large rising balloons. These instruments transmit accurate temperature measurements (and other data) back to base by radio transmissions.

  Since the mid-1970s, weather satellites have added to the volume of temperature measurements by carrying instruments that measure atmospheric temperature in a different way. By downward probing with infrared or microwave sounding beams from a height of 850km or so, the satellite instruments are able to convert their measurements to temperature values through the depth of the atmosphere and also on the surface of the Earth.

Although less accurate, the satellite instruments are able to provide a great number of atmospheric temperature values, many more than the number provided by the traditional balloon measurements. This satellite data is very important to meteorology and it is included in the global warming studies.[1]

In recent years dissenters have shown that the satellite-derived temperature records of 1978-2004 do not accord with the warming record of other measured data; instead, the satellite measurements indicate much less warming. So the dissenters insisted that the global warming predictions were considerably overdone. This disparity in the measurements proved to be very puzzling for some time and, at first, the satellite instruments were suspected of inaccuracies. However an explanation has since emerged.

Unknown to the collectors of the satellite data, the orbit of the satellite had changed very slightly. This had been caused by the effect of the very small amount of atmospheric friction that still exists 850km above the Earth, where the satellite was orbiting. The friction had acted upon the satellite, causing it to slow down a fraction and, as a result, the vehicle moved into a slightly lower orbit. This in turn upset the timing of the satellite measurements; for instance, a satellite temperature value thought to have been taken at midday was actually skewed to a much later time in the day when the temperature was lower. Corrections have now been applied and the satellite and traditional observations now accord.

– *Reliability of historical data questioned:* there have also been arguments about the reliability of the historical temperature data and the means of their manipulation. Recently (2009-10), there have also been claims of processing errors in the databases. This has led to a UK proposal that there should be a new international analysis of all three major world temperature databases; this has been agreed internationally and is expected to take place in the next few years. However the majority of scien-

---

1 Appendix B: Weather Satellites see p. 218

tists stress that this process of reorganisation and, if necessary, correction, is not expected to alter the fact that world temperatures increased during the 20th century and will continue to do so.

Dissent in the Bible: interpretation and truth

There is a lot of dissent and disobedience in the Bible and it is roundly condemned as a vice! Dissent is identified as the opposite of obedience, which is always regarded as a virtue. Christians believe that the Bible contains God's truth and so there must not be any attempt to twist its words, especially for personal gain.

There are many texts praising obedience and condemning dissent:[1] however the dissent and obedience found in the Bible usually refers to the people's relationship with God; God the Creator wants his people to be obedient to his perfect plan and often condemns any occasion where this does not occur. In the New Testament, there are a number of occasions when Jesus shows absolute obedience to God; this is at its most extreme when he accepts intense suffering and death on the Cross in obedience to God.

However it is accepted that there are many places in the Bible where interpretation is required and it is obvious there may be a number of interpretations applied to the same biblical passage. Also, interpretations will change over time as a result of new information and studies. This has been the situation down the centuries since the Bible was written. Most Christians find that these variations and changes are acceptable if they are offered in a spirit of love and truth-seeking. However it is unacceptable to dissent from the clear teachings of God or Jesus Christ, especially if this is done for selfish or dishonest purposes.

---

1 e.g. obedience and dissent: Proverbs 15:26-28; obedience: John 6:38; dissent: 1 Corinthians 1:10-13.

Although the Bible does not provide any specific texts that fit directly with present day scientific argument and dissent, the biblical concept is nevertheless applicable to modern scientific methods. As in all parts of life, the scientist should formulate his arguments from purity and truth. In other words, the arguments advanced should be valid, based on scientific method and properly researched. They should not be advanced on the basis of unscientific method, prejudice or self-interest – for these do not represent truth.

The extent of the human contribution to global warming

The links between energy, fossil fuels and carbon dioxide

The major fossil fuels comprise oil, natural gas and coal, all of which are used worldwide to generate energy. Coal is the earliest of these fossil fuels; it has been mined as a fuel source for a very long time, stretching back 10,000 years or so. Of course wood burning was an even earlier energy source but wood is not a 'fossil fuel' by definition; however the process of burning wood to release energy emits smoke filled with gases and solids of various types.

Wood is still burnt as a fuel but its use tends to be greatest in areas outside the developed world. In the process of energy creation, the burning of fossil fuels creates greenhouse gases, especially carbon dioxide. Apart from water vapour, carbon dioxide is the greatest contributor to the greenhouse effect.

It is estimated that fossil fuels provide around 80% of the energy that is consumed by humans worldwide. Nearly 40% of this is provided by oil with the remaining 40% split almost equally between natural gas and coal. The 20% balance of energy not generated by fossil fuels is supplied largely by nuclear fuels and hydro electricity with very much smaller contributions from other alternative energy sources. Nowadays, there is a great push to increase alternative energy sources. It has been estimated that worldwide energy requirements are growing by almost 2% per year; however, against the background of rapid industrial development in the

Asian and African continents, this growth figure may need to revised upwards even further.

Fossil fuels are non-renewable

Aside from their polluting effects, one of the major characteristics of fossil fuels is that they are a non-renewable energy source. The financial analogy sometimes used is 'spending savings from the bank'. Furthermore, it is the rapid spending of a resource which has literally taken millions of years to make. This means that there is no question of acquiring more stocks of natural fossil fuel materials; at some time in the future, fossil fuels will run out and will no longer be available to produce energy.

It is difficult to estimate 'run out' times because the full extent of fossil fuel resources cannot be known precisely. Also, the cost of producing these resources comes into the equation; there may come a time when it is too difficult, dangerous or expensive to extract a particular fossil fuel reserve. That said, the run-out times for gas and oil are usually suggested to be in the time frame of 50-100 years from now, with coal rather longer at 200 years at least.

Of course this implies that the global warming generated by burning fossil fuels will eventually diminish and stop within several centuries from now. Does this mean we could just leave the problem to solve itself in time? The IPCC reports and predictions show that the answer to this question must be 'no'. Flooding by sea level rise is already a serious problem, along with unfortunate ecological and animal behaviour changes. More severe weather events are predicted to be 'very likely' and Chapter 5 will show how these give rise to detrimental implications for all.

Even catastrophic climate change, arguably the most apocalyptic of the predicted events, has been shown to be sensitive to the higher levels of global warming which would occur if unchecked global warming occurred in the coming decades of this present century. Therefore we need action now to mitigate the global warming which will occur during this century and next while fossil fuel resources are still in full use.

The evidence for blaming us

The IPCC scientists base their conclusions on that latest and most sophisticated climate modelling experiments which were described earlier in this chapter. Although there have been some accuracy problems associated with world historical databases, the conclusions of the scientists remain intact. In particular they remain confident that the correction and refinement processes which will be applied to the databases will not alter their essential advice – which is that the world has been warming significantly during the 20th century and more especially in the last 50 years.

When the computer experiments were done, the simulation which combined natural warming factors plus human contributions from 1960 to the present day replicated with good agreement the actual global average temperature record during that period. This was accepted for the IPCC 2007 reports and provides a valid scientific case to support the claim that humanity has been and is a major contributor to global warming.

Of course we are not *totally* to blame but the evidence shows that our contribution to global warming since 1960 has caused the temperature rise to be increasingly steep. Furthermore, it seems reasonable to suggest that our continuing contribution will push global temperatures higher and higher. At present, our pollution has resulted in an increasing concentration of greenhouse gases caused mainly by the burning of fossil fuels for energy creation.

The consequence remains a question of atmospheric balance; our additions to greenhouse gas accumulations have been sufficient to overload the natural compensating systems. Without our contribution, the experiments have indicated that the global temperature rise may well have steadied or, at least, risen much more slowly.

Unfortunately, the scientific evidence makes it pretty clear why we're being blamed!

4

Unfortunately There Is More...

It has become commonplace for global warming to be identified as the cause of all aspects of weather and climate change. However there are several other mechanisms which must be considered. Over the period of global warming, it will be seen that these other mechanisms have also contributed to the overall problem.

Industrial pollution and acid rain... 1950s

Although the title suggests this problem was linked to the 1950s, serious industrial pollution and acid rain actually started two centuries before, in the mid-18th century. This was the time of the Industrial Revolution when heavy industry in many countries started generating more smoke and chemical pollution than ever before. The term 'acid rain' began to be used in the 19th century but the problem was not focussed upon scientifically until the 1950s.

The work that was done then led to an understanding of the effects of acid rain, not only in plant damage and destruction but also in how it affects the wider eco-systems of the world. A positive outcome has been a range of international agreements to try to address the problems identified.

What is acid rain?

Acid rain is atmospheric precipitation with significant concentrations of sulphuric or nitric acids within the water drops. Frozen precipitation such as sleet, snow or hail may be similarly contaminated. In addition the term 'acid rain' also covers dry deposition; this happens when acidic dry particles are carried aloft in the air and subsequently deposited on the ground. Both sorts of acid rain are produced when the low-level atmosphere is polluted by sulphur dioxide and nitrogen oxides. These chemicals are emitted by many industrial processes, including those that burn fossil fuels, and by all types of vehicles powered by petrochemicals.

The degree of acidity of acid rain is measured by the PH scale which has a range from zero to 14.0, lower numbers being increasingly acidic. Pure water is neither acidic nor alkaline and takes the central PH value of 7.0. In the 'wet' form of acid rain, the polluted air rises through the clouds and the acid pollutants combine with water vapour to form sulphuric and nitric acid droplets. When precipitation falls from such clouds, the raindrops are significantly acidic, with PH values of 5.5 or less. If the ground upon which they fall has also been polluted by dry deposition, the degree of acidity is increased even further. Acid rain with PH values as low as 2.0 have been measured around some very highly industrialised areas.

Acid rain is a serious problem in the vicinity of heavy pollution sources; the very worst effects are found in areas immediately downwind of these sources. Upper wind statistics reveal where the most affected downwind areas are likely to be. World maps of acid rain identify that high acidity pollution is a feature of the northeastern parts of North America, much of northern Europe and all the heavily populated parts of eastern Asia. Every one of these regions has significant industrial output.

What does acid rain do?

It harms vegetation, including all kinds of plants, trees and agricultural crops. The acid interferes with the plant's ability to acquire nitrogen nutrients, an essential process called 'nitrogen fixing', and also acts to remove existing nutrients from foliage. Also, plant germination processes are disrupted. Thus the plant's ability to function and maintain its health is impaired. This impinges directly on humanity because plants are part of the cycle of life that involves the world's animal life. Plant life (of all types) are the major producers of the oxygen we breathe; they also remove carbon dioxide from the air.

Aquatic life is particularly adversely affected by acid rain. The environment of lakes and rivers whose water has a PH value of 6.5 or so are filled with a variety of healthy life – insects, fish, birds and other aquatic animals; all these lives interact and are generally dependent on each other. When the water becomes increasingly acidic as a result of direct acid rain pollution and/or acidic rainwater flowing down from higher ground, there is a marked decrease in aquatic life activity. When the PH value reaches 4.5, the water becomes dead to life.

Furthermore, acid rain is corrosive, attacking and damaging the metalwork and stonework of buildings and other structures. This is a serious problem for the preservation of historic buildings, bridges, etc.; it also boosts considerably the maintenance and repair costs for all these structures. Obviously, all other buildings and structures are adversely affected too.

Significantly, the problem of acid rain worsened between 1950 and 1980. The reasons have their origin in the same circumstances that increased greenhouse gases. Firstly, the population of the world was expanding rapidly and affluence was increasing. As a result, much more energy was being consumed. The additional energy required the burning of increasing amounts of fossil fuels in industry, production and transportation. The increase in effluents provided more acidifying agents and acid rain pollution was boosted markedly.

The general increase in acid rain pollution resulted in a wider spread of its effects. The greater concentrations of damaging pollutants were carried to higher levels of the atmosphere and transported considerable distances by the powerful upper winds which are found there before being deposited in areas hitherto unaffected.

What solutions have been applied since the 1980s?

Although the long-standing problem of acid rain was recognised scientifically by the mid-20th century, it was not until the 1980s that international solutions were negotiated. There were two basic reasons for this. Firstly, research had shown clearly that the problem of acid rain was worsening. Secondly, this was the time when a number of countries complained bitterly that they were being deluged with highly acidic rain pollution from one or more industrial neighbours upstream! Countries particularly affected were Canada and the Scandinavian countries of northern Europe.

This led to a great deal of consultation and eventually various important agreements, protocols and acts were signed in North America and Europe. This has had a positive impact, because the past decade has seen a decrease in acid rain in these regions. However it is pointed out by those who monitor such effects that reversing ecological change takes a considerable time.

In the same decades, great concern was expressed about the emissions from eastern Asia. This has led to cooperation agreements between a number of countries in these regions. A network of acid rain monitoring stations has been set up. A number of eastern Asia countries has acted to reduce their emissions of nitrates and sulphur dioxide, with positive effect; however the rapidly expanding industrialisation of these regions continues to create significant acid rain problems.

Concern is also expressed about increasing industrialisation in areas currently unaffected by the acid rain problem. In the developing world, rapid and often uncontrolled industrial emissions have been observed from both existing and new developments. It is feared that acid rain will become a much increased hazard in these areas and that this will impinge upon the global problem.

Acid rain and methane emissions

In recent years it has been reported that there is a link between acid rain and methane emissions. Statements like 'Acid rain reduces global warming' have been published by the media. Could this be a mitigating factor in the global warming problem?

Most atmospheric methane comes from totally natural sources as a result of vegetation decomposition in marsh-type environments. Farm animals also contribute very significantly to methane production. However the considerable waste produced by the human race tends to be deposited in landfill sites and this also contributes methane to the atmosphere as it decomposes. The decomposition process involves methane-emitting microbes and the gas they produce is a large proportion of the methane found in the atmosphere. When acid rain falls on wetlands and landfill sites, it has been noted that the activity of these microbes is reduced and, as a result, less methane is released. In theory, therefore, less methane means less greenhouse effect.

Although this effect of acid rain on methane production tended to be presented as a new discovery, it has in fact been known about for several decades. Today's extremely powerful computer systems permit very sophisticated climate models to be run and the effect of acid rain on methane emissions is one of the many factors that can now be included. Results from these simulations indicate that methane is not likely to be a major contributor to the greenhouse effect; it is thought that even a large reduction in methane emissions would still have a very small overall effect on global warming.

The Bible urges responsible stewardship of the World

The story of acid rain over the last few decades is one of at least partial success. Countries and regions who were responsible for significant acid rain pollution downstream have acted to mitigate the problem; however there is concern about certain areas where rapid industrialisation is taking place.

The Bible is unequivocal about man's stewardship responsibilities. They start from Adam's responsibility for the Garden of Eden[1] (which is intended to represent the world) and all its plants

---

1 Genesis 1:26-29

and animals, leading to God's command to 'Love your neighbour'[1] and then to the prediction in the book of Revelation '...the time has come for rewarding... those who reverence your name.[2] Stewardship means responsible and conserving action, not using the resources of the Earth in selfish and casual ways. The Bible insists that only those who obey God's stewardship responsibilities will receive God's promises set out in the book of Revelation.

Upper atmosphere pollution by aviation... 1960s

During the 1960s and 1970s, there was significant concern about the pollution of the lower stratosphere by high-flying aircraft. In earlier decades, civil and military aircraft were powered by piston engines driving propellers. This type of aircraft flew at significantly lower levels, flying almost invariably in the troposphere. The troposphere is the part of the atmosphere that we live in, extending from the ground upwards for many thousands of feet.[3] However from the 1960s, both civil and military aircraft began to be powered by jet engines, which were designed to fly most efficiently in very cold temperatures. As a result, aircraft began to fly routinely at stratospheric altitudes. This gave rise to the fear that this new source of stratospheric pollution would lead to significant 'stratospheric warming' and that this could have serious implications for the climate of the world.

It had been known for decades that the upper atmosphere is subject to a natural phenomenon called 'Sudden Stratospheric Warming' (SSW); these warming effects are observed to occur cyclically and are of various strengths from year to year. After much research, it was possible to explain all the observed stratospheric warming events within natural global cycles and eventually it was concluded that the effect of high-flying aircraft was insignificant.

1 Luke 10:27
2 Revelation 11:18
3 Appendix B: Troposphere and Stratosphere see p. 217

It is always wise to remember that mankind (even scientists!) is fallible. In the Bible, God accepts the fallibility of man; this is why he makes it possible for our sins to be forgiven. The Bible confirms this act of love many times.[1]

Global dimming… 1970s

Global dimming is not generally well known. Unlike global warming, global dimming is a term describing a phenomenon that has only relatively recently been accepted as scientifically valid. Yet it seems that this latest contribution to the worldwide weather modification discussion is of significant importance, especially if one listens to the scientists who have been working on the problem. The latest evidence provides a powerful case for the existence of global dimming and for its importance as one element in the weather and climate modification scenario.

The phenomenon was first noticed several decades ago when climatologists working in the Middle East noticed a decrease in surface water evaporation when compared to the records of the previous 10 or 20 years. For a considerable time and in many locations all over the world, the amount of water evaporated into the atmosphere has been recorded each day by measuring how much water had to be added to refill a large open pan of a standard size. Obviously, the more water that has to be added, the higher the evaporation rate has been for the period since the last top-up. Unsurprisingly, such measurements are referred to as the 'pan evaporation rate'.

In the 1970s, decreased pan evaporation rates were recorded and investigation suggested that a reduction in solar radiation was responsible. The 10-20 year decreases in solar radiation were not constant over the globe but varied considerably from place to place, amounting to as much as minus 30% in some places. Further measurements suggested that the decrease of solar radiation was proceeding at around 0.3% per annum on average.

---

1 e.g. Acts 3:19

The scientists involved in this research published their results but their findings were decried and ignored.

The reason for this reception was simple. The scientific world pointed out that global temperatures were not falling - they were rising! Global warming had already been identified and scientists were measuring small but significant temperature rises across the planet. Also, the scientists who studied solar energy levels reported that there were no significant changes in the level of emissions from the sun.

However opinions began to change when the measured decrease in solar radiation received on Earth was confirmed from other sources. In an experiment on the Maldive Islands in the Indian Ocean, it was noted that the more northerly islands in the group received significantly lower levels of solar radiation than the islands in the south; the solar radiation difference was in the order of 10%.

Further study revealed the more northerly islands to be affected by significant atmospheric pollution (large particles of smoke, soot, etc.) produced by distant industrial and urban sources while the more southerly islands were largely unaffected. This explained the anomaly and established a positive link between solar radiation levels and pollution caused by large amounts of carbon and other solid particles in the atmosphere.

Strong confirmation of these results came from a completely separate observation in North America. Following the 2001 '9/11' terrorist attacks in New York and Washington, commercial aviation over the whole of the USA was grounded for 3 days. As a result, the skies over the whole country were not subjected to the carbon and other pollutants emitted by civil aviation aircraft engines.

Importantly, the atmosphere was also cleared of aircraft 'contrails' (condensation trails). These are the common and characteristic streams of cloud that are often trailed from aircraft flying in the high atmosphere. Depending upon the condition of the atmosphere, contrails may dissipate quickly but they are often remarkably persistent and sometimes spread out to become larger areas of persistent cloud. Obviously, contrails obscure the sun to some degree.

The experiment conducted on these unique days found that that the removal of civil aviation pollution and contrail clouds coincided with a sudden increase in the air temperature over North America - the actual value was just over 1 DegC. This may not sound too dramatic but +1 DegC in 3 days is actually the most rapid temperature rise in world history! This sudden jump in temperature may also be compared with the +0.75 DegC recorded for global warming over the past century.

How does global dimming work?

The pollution that causes the global dimming effect is not associated directly with the greenhouse gases that cause global warming. Global dimming pollution consists of large particles of solid carbon, soot, etc. This is the sort of pollution that is generated near the Earth's surface by almost all processes of energy use, especially when equipment is old or badly maintained. Typical sources are factory and domestic chimneys and the engine exhausts of older vehicles. This type of pollution is the dark haze that can be seen floating in the vicinity of many industrial and heavily populated regions. In addition, aircraft make their contribution from the air high above the Earth, expelling solid carbon products directly into the high atmosphere as well as the water required to make contrails.

While pollution by carbon products alone has some effect on the sun's radiation level, it is the combination of carbon particles and water that provides the mechanism for global dimming. The large carbon particles are carried upwards by rising air currents and through cloud formations. Clouds are made up of very small water droplets; each droplet needs to condense around a minute particle, called a 'condensation nucleus'. These are often tiny dust particles, sea salt or pollen grains. Then, when the water droplets are subject to precipitation formation processes, raindrops may fall from the cloud.

However when much larger particles of carbon are introduced into the cloud, normal processes are disrupted and the water vapour condenses *upon*, not *around* these larger particles. As a result, a number of water droplets will cling to one large carbon particle. When these large particles are carried to the top of the cloud, the effect is to present a carpet of very highly reflective water droplets. This reflects back into Space much more solar radiation than a non-polluted cloud would; consequently, the solar radiation that penetrates to the ground below a carbon-contaminated cloud is significantly reduced.

The consequences of global dimming

Quite simply, global dimming reduces the temperature in the atmosphere below the cloud. So it is clear that global warming and global dimming act against each other. Global dimming restricts the temperature rise associated with global warming. If global warming was not occurring, the mean temperatures of some areas of Earth would be decreasing because of global dimming. On the other hand, if global dimming was reduced or removed, a higher rate of global warming would result.

Since the recognition of global dimming, scientists have attempted to assess what the direct effects of global dimming are. This work has indicated that some of the climate disasters of the part 30 years are likely to have their origin in global dimming. In particular, the reduction of solar radiation reaching the ground and the lower atmosphere has been sufficient to upset equatorial weather patterns.

The 'Inter-tropical Convergence Zone' (ITCZ) is a meteorological feature that is generated in equatorial regions, associated with converging winds at low levels and large-scale upward vertical motion. This produces a wide belt of thick, often convective cloud from which heavy rain falls. Thunderstorm activity is common. The ITCZ is usually very visible on 'full disk' satellite[1] pictures as a wide but ragged band of bright cloud extending all round the world near the Equator.

During the year, this large belt of stormy weather oscillates north-south around the Equator, delivering life-giving rain in the equatorial southern hemisphere during their summer season and transferring to the northern hemisphere when the sun 'moves' there for the northern hemisphere summer season. The usual name for this season of rain is 'monsoon', often prefixed by a regional name – e.g. the African Monsoon.

Studies have revealed that reduction of solar radiation by global dimming was sufficient to disrupt the normal movement of the ITCZ; in consequence, its belt of life-giving rain did not transfer to its normal summer position in the northern hemisphere. The result was that certain regions of Africa (for instance the normally fertile Sahel area) remained dry and crops failed. This happened frequently in the 1970s with catastrophic effects on the populations; at least 50 million people perished in that period and there were huge losses of farm animals and agricultural land.

In such circumstances, it is not long before the process of desertification takes over. This is partly to do with the prolonged dryness and the death of plant life but more importantly with the local population's abandonment of the area; with the land producing nothing but withered crops and dying animals, the community gives up and the desert is quick to take over. Once abandoned to the desert, it is extremely difficult to reinstate the land.

History has provided many examples of desertification. In the times of the ancient Roman Empire, North Africa was a fertile and forested land, known to the Romans as 'the breadbasket of Rome'; it has been thought that a process of deforestation was responsible for its change of status over the centuries. Similar processes have taken place in the mid-west of the USA with fertile land turned into a 'dust bowl' by incorrect farming methods and the subsequent removal of the topsoil layer by unseasonably strong winds. This has also happened across the once fertile plains of Russia.

---

1 Appendix B: Weather satellites, see p. 218

Regarding the disruption of the ITCZ, there is some encouraging news to report. Recent years have shown some decrease in pollution levels, especially from Europe, and it appears that this has impinged positively on the situation. Happily, there has been a reinstatement of more normal ITCZ movements, with rains returning to more normal patterns.

The realisation that global dimming acts against global warming has led some people to a tempting thought: 'Could global dimming be increased to balance global warming and thus stabilise world temperatures?' While it is undoubtedly true that global warming and global dimming act against each other, the thought of the world's governments and populations working hard to cause greater pollution is both ridiculous and unsettling. It is also highly dangerous, because the effects of more global dimming will of themselves impose unknown consequences on the world's climate, of the sort already demonstrated in equatorial regions, or even worse.

The 'darkened suns' of the Bible

The darkening of sun, moon and stars features in the Bible a number of times. In the Old Testament, this is often linked with the sin and evil acts of God's people or with those who oppress them; in particular, the darkening is linked with the 'Day of the Lord', a day of retribution for the evil-doers. These warnings are found twice in the book of Isaiah and in Joel.[1]

The texts from Isaiah are repeated in the New Testament but are used for a different purpose. Here the link is with the End Time when Jesus will return. In both Matthew and Mark, the darkening of the sun is associated with the distressing events which are predicted to occur before the return of Jesus.[2] Darkening of the sun also appears in the predictions of the book of Revelation. In this account, the darkening is caused by dense smoke being

1 Isaiah 5:30, Isaiah 13:10, Joel 3:15
2 Matthew 24:29, Mark 13:24

released from the Abyss - the Abyss represents an opening to the Underworld, the place which contains the power of evil.[1]

Probably the most well known darkening of the sun was associated with the Crucifixion of Jesus. The gospels told us that Jesus was crucified at the 'third hour'; the Jewish day starts at 6am so the time of crucifixion was 9am. The gospels recorded that from the sixth to the ninth hour 'there was darkness all over the land'. [2] Therefore the reported darkness occurred from midday until 3pm when Jesus died. Although there have been a number of attempts to explain the darkness logically and even scientifically, none of these are plausible. Instead, the darkness at the crucifixion is reasoned to be a symbol of Jesus' separation from God, essential so that his physical death was possible.

Certainly all the occurrences of darkened suns in the Bible are associated with negativity and suffering: this accords with the effect of global dimming which has caused the destruction of large tracts of formerly fertile land in the equatorial regions, imposing suffering and death on the people and animals who lived there.

Stewardship, responsibility, compassion? Some progress...

As stated earlier in this book, the Bible calls for stewardship of the Earth. God calls us all to be careful and responsible in our treatment and usage of the land and sea. He calls for us to be humane and caring of all plants and living creatures. He calls us to be compassionate to all our fellow human beings. Jesus calls for us to love one another as he loved us.[3]

Since the reason for global dimming has been determined, there has been some corrective action. As a result the ITCZ has re-established a more normal pattern and tropical rains have fallen in their 'proper' seasons. This is a good result and is in accordance with what the Bible teaches. Indeed the Bible states several times that the reward for obedience to God's stewardship requirement

---

1 Revelation 9:2
2 Matthew 27:45, Mark 15:33, Luke 23:44
3 John 13:34

will be the delivery 'correct' seasonal weather, in particular, rain when it is needed for the sustenance of plant and animal life.[1]

On the other hand, the Bible teaches that disobedience will have the opposite effect![2] There is still considerable concern that unregulated industrial development in developing areas of the world will maintain global dimming or make it worse. Acceptance of world stewardship responsibilities is required there. Where possible, the richer nations of the world should assist the poorer developing nations to embrace world stewardship, for the benefit of all.

The Ozone Problem ... 1970s

Ozone is a form of oxygen. The common form of oxygen is of course the gas that all animals need to breathe; this has two atoms per molecule. Ozone differs from this with three atoms in each molecule and is therefore a completely different gas. Pale blue in colour, it has a characteristic sharp smell and is toxic and explosive, even in small concentrations.

Ozone occurs naturally in small amounts in the stratosphere where it performs the important role of helping to absorb damaging ultraviolet (UV) solar radiation. The absorbing action of ozone protects many organisms living on Earth from serious physical damage. In particular, it has long been established that an excess of UV radiation causes skin cancers. It is also damaging to eyesight and to the immune system of the body.

The concentration of stratospheric ozone over the Earth's South Pole varies on a yearly cycle. Each year, ozone is depleted in the southern hemisphere spring. This is a consequence of the total absence of the sun during the Antarctic winter which results in very low atmospheric temperatures above the polar region. Ozone concentrations have been manually recorded for many years and

---

1 e.g. Deuteronomy 11:14, Zechariah 10:1, Acts 14:17
2 e.g. Deuteronomy 11:17, 2 Samuel 1:21, Isaiah 5:6

since the 1980s there has also been direct satellite temperature monitoring.

The ozone depletion over Antarctica lasts around two months each year and then the area of low ozone concentration breaks away to traverse some of the most southern land areas of the world (the Falkland Islands and the southern tip of South America). At such times, UV radiation levels are greatly increased and this brings a serious health hazard to the people who live in these areas.

The 'Ozone Hole'

In the 1970s, Antarctic scientists noted with concern that the concentration of ozone over the South Pole region had decreased even more dramatically than usual. In fact the decrease was so great that the recordings were doubted at first. This was the phenomenon that became known as the 'ozone hole'. These observations caused great concern because of the implied, very serious health dangers

More recently, an ozone hole has been identified over the North Pole too. Here, the depletion is not so marked and the ozone hole is a good deal smaller. This is because the North Pole region (with no land mass) does not experience temperatures as low as the South Pole. As expected, ozone depletion in the Arctic occurs in the early spring, that is, February and March in the northern hemisphere. Like the Antarctic ozone hole, the Arctic ozone hole is potentially dangerous to the populations of the most northern latitudes.

Ozone concentrations and their variations are not confined to polar regions, though it is here that the changes are most marked. There are ozone concentrations in the stratosphere all over the world. In the last decades these concentrations have also shown a decrease of 5-10%. This has some significant implications for world health.

Investigations started in the 1970s soon revealed that the release of CFCs (identified as greenhouse gases in Chapter 2) had caused a very serious environmental problem. CFCs were apparently harmless chemical compounds first manufactured in the 1930s; they were used as propellants for aerosol cans, in refrigeration and in various other manufacturing processes.

However it was discovered that CFCs released near the surface of the Earth accumulate in the stratosphere where UV radiation breaks them down chemically. The released chlorine atoms then react with ozone and destroy its protecting role. Worryingly, the investigations emphasised that very small amounts of CFCs have a huge effect, especially in very cold polar regions.

The seriousness of this situation was recognised internationally and even a decade ago most countries had signed agreements to stop production of CFCs and similar ozone-harming chemicals. It is now generally accepted that this initiative has been successful since the concentration of ozone destroying chemicals in the atmosphere is dropping. On the other hand, very large ozone holes are still forming over Antarctica.

There has been great variability in Antarctica ozone holes; an extremely large hole developed in the year 2000 (over 28 million sq km, noted to be of record size); by contrast, the 2002 hole was much smaller. Subsequent years have seen a return to larger ozone holes, with the 2006 ozone hole equalling the year 2000 extreme.

Some scientists now believe that the ozone hole situation is in a recovery period but others point out that the CFCs and their elements are very stable substances that will continue to affect the ozone situation for decades to come. The current suggestion is that the ozone hole may be truly repaired by the mid-21st century when ozone fluctuations will return to more normal levels.

Ozone and global warming: some success...

To add to the problem, ozone is also a powerful greenhouse gas. It may therefore have something to contribute to the process of global warming. The restoration of stratospheric ozone levels may largely solve the dangers of health hazard but have the potential to add to the global warming process to some degree. This effect has been studied and, luckily, the low concentrations of ozone would appear to make it a very minor player in the totality of global warming. Nevertheless it is appropriate to keep ozone in mind when considering all the global warming factors.

It is claimed that global actions on the ozone hole problem have been timely and successful, although monitoring needs to continue. International action to outlaw CFCs and similar gases have diminished (but not removed) a serious health hazard. In some senses, the ozone problem was easier to address; its cause and effect was established quite quickly and its effect was directly towards the health of our fellow man. Global warming, global dimming and acid rain are arguably more difficult, diffuse and unpredictable.

Certainly, a focus of stewardship has been demonstrated and this is to be celebrated. Inevitably, there are still those who manufacture CFCs illegally or who produce new gas compounds that produce a similar effect. Rightly, this is censured and prosecuted by law if possible.

5

'Extreme Weather Events'

The IPCC warns us that 'extreme weather events' are likely to become more frequent in the future. Currently, this is not expressed as a certainty but with a confidence level of 'very likely'. During the last decade, this particular warning has been strengthened progressively. Some scientists claim that extreme weather events are already happening across the world; certainly, the media tends to categorise every occasion of bad weather as a demonstration of global warming and climate change!

What is covered by the term?

Most people tend to think of 'extreme weather events' as violent storms, powerful winds, heavy and prolonged rain, deep snow, thunderstorms, tornados and the like. In other words, episodes of rather violent weather. However the IPCC warning covers other sorts of extreme weather types, like extended periods of heat or cold, persistent fog, ice or concentrated pollution in the low atmosphere. All these have the potential to create serious problems. Then there are the consequentials of some types of extreme weather – for instance flooding after torrential rain or physical damage caused by the effects of freezing or drought. The consequentials can easily be as serious and destructive as the weather that causes them; furthermore they can affect areas remote from the causal weather.

Tropical cyclones

Tropical cyclones (also called hurricanes, typhoons and various other regional names) present extreme weather in its most violent mood. The most powerful of these huge low pressure weather systems cover thousands of square kilometres (sq.km) and, as they move, subject huge swathes of sea and land to violent winds and torrential rain. As their generic name suggests, this type of disturbance forms invariably in tropical regions, often near 20 Deg N or 20 Deg S but the rapidly-spinning cyclone may then move to higher latitudes before weakening over the cooler seas or land areas it finds there.[1]

However, even when tropical cyclones move to higher (cooler) latitudes and weaken to become 'extra-tropical', they are likely to remain areas of severe weather for some time. Also relevant is the claim that tropical storms are becoming more frequent, that is, the number of cyclones during the tropical cyclone season is increasing. This suggests an increased chance of disaster to the at-risk regions.

Meanwhile, those who live in higher latitudes are beginning to ask whether the change to 'more extreme weather' will cause tropical cyclones to affect areas much further from the Equator, perhaps at 40, 50 or even 60 DegN or DegS? The usual answer is 'no', because higher latitudes do not (and will not) develop the meteorological conditions required for tropical cyclone generation or maintenance. However a continued rise in sea temperature implies that tropical cyclones could form slightly further north or south of the Equator and have the potential for a somewhat longer life than they have now. So there is suggestion of increased danger to fringe latitudes.

---

1 See Appendix B: Tropical Cyclones, see p. 216

It is also true that some higher latitude low pressure systems can have their origins in a tropical storm circulation; these circulations can sometimes redevelop to produce severe storms in mid-latitude regions. It could be argued that the greater incidence of tropical storms suggested in the last paragraph could be responsible for more deep low pressure systems in mid-latitudes. However this link is tenuous because low pressure formation and maintenance in higher latitude regions requires many other development factors to be present, most of which are unrelated to tropical features.

Mid-latitude depressions (low pressure systems)

These are cyclone systems that affect a wide band of mid-latitudes, mainly within 30-70 Deg N or Deg S. The systems may vary greatly in size, everything from minor low pressure developments associated with an area of cloud and light precipitation to very large and active cyclones accompanied by huge amounts of precipitation and gale force winds. Such large cyclones can be seen on satellite imagery as very large areas of deep cloud. In some of the most developed systems, clear spirals of cloud are sometimes visible.

The most active of these mid-latitude depressions can certainly produce extreme weather conditions, sometimes approaching the weather intensity of weaker tropical cyclones. Surface winds in the huge circulations can be extremely strong and rainfall is often intense and extensive. Depending upon position and temperature structure, many precipitation types can be generated; very heavy rain, sleet, snow, hail and thunderstorms all occur.

The IPCC prediction of 'more extreme weather events' can certainly be delivered by mid-latitude depressions. The development and intensity of such systems is usually determined by temperature discontinuity; the hotter the warm air in the system and the colder the cold air means increasingly strong development. Global warming means that higher temperatures will be available and system development stronger, in turn producing more violent winds and even heavier and more prolonged precipitation.

Weather fronts

These are very important features on mid-latitude weather maps. A weather front is the boundary between two air masses which have different origins and characteristics. Weather fronts are areas of atmospheric uplift which normally generate cloud and precipitation. They are very variable in intensity; the most intense are capable of producing very strong winds and a great deal of precipitation, including rain, sleet, snow, hail and thunderstorms, while a weak weather front may only be marked by a fragmentary band of cloud.

Meteorology defines several types of weather fronts. A warm front is a boundary which moves in such a way that warm air replaces cold air. On weather maps it is usually shown as a red line. Because the warm air is less dense than the cold air, it rises up over the cold air, often at quite a gentle rate. This is why warm front precipitation is often (but not always!) quite light.

A cold front acts quite differently. Here, the movement of the frontal boundary replaces warm air with cold air. Cold fronts are usually shown as a blue line on weather maps. This time, the cold air is more dense and it undercuts the warm air, causing it to be lifted from the Earth's surface quite quickly. These stronger upcurrents encourage the precipitation associated with a cold front, generally sharper and more intense.

A third type of front is called an occlusion. This is formed when a warm and cold front are brought together – normally a cold front catches up with a slower moving warm front. Unsurprisingly, an occluded front, being a combination of a warm and a cold front is usually marked on a weather map as a purple line. Precipitation from an occlusion is more variable and will usually depend upon the temperature differences between the air masses involved.

The intensity of fronts is normally determined by the temperature change across the frontal boundary; the greater the temperature change, the more intense the front and the more severe the associated weather. Global warming suggests that frontal temperature discontinuities will be increased; stronger wind and more intense precipitation will result.

Heavy showers

By their nature, showers are more local features, generally more short-lived, often because the shower cloud moves quickly across an area, driven by the winds at cloud level. Shower precipitation is very variable, everything from a few spots of light rain to a drenching downpour. However the mechanisms that cause heavy showers can produce quite violent weather, with intense precipitation and thunderstorms. Showers may become concentrated into an area or a line and this has the potential to intensify and prolong the effects.[1]

Again, the higher temperatures produced by global warming have the potential to make showers heavier and more prolonged with enhanced gusty winds.

Persistence extremes

The other types of extreme weather take their severity from persistence rather than the direct violence of their effects. Short periods of heat, cold, fog or ice, etc. can be tolerated by most people and infrastructures but when these extremes persist, significant problems often develop. Persistent weather extremes are usually associated with settled, unchanging weather conditions which are a product of largely static weather patterns.

Typically, a large high pressure area dominates a region for an extended period. An important characteristic of high pressure is that the air in the huge circulation sinks towards the ground and becomes hotter. Eventually, a marked temperature inversion may form only a few thousand feet above ground and this acts to trap smoke and other types of pollution. This pollution becomes more dense building downwards towards the ground.

With persistence extremes, the involvement of global warming is rather more subtle. Air movements all over the world are held in balance, with upward air movements balanced by downward

1 Appendix B: Rain and showers, see p. 210

movements elsewhere. Upward air movements are generally associated with 'bad' weather, that is, cloud, precipitation and strong surface winds. This is the sort of weather that is associated with tropical storms and also with powerful mid-latitude depressions. The intensity of such systems is determined by high temperatures which provide the energy for the upward movements of air; then, a plentiful supply of moisture (often from a relatively warm sea) produces deep clouds and heavy precipitation.

So if global warming encourages the formation and maintenance of more intense tropical storms or mid-latitude depressions, it follows that the balancing downward motion of air in another part of the world needs to be stronger and/or more extensive. This, then, is a mechanism for making such areas of high pressure more intense, persistent and extensive. In turn, when persistent extremes become associated with these large high pressure systems, they are more intense and long-lasting.

– *Persistent low temperatures:* large and static high pressure weather systems commonly cause extended periods of cloudless skies and light winds. If the season is winter, especially in higher latitudes, such conditions will drive down air and ground temperatures so that sub-zero temperatures persist not only at night but throughout daytime, too. Such low temperatures bring their own problems but will often be associated with icy conditions underfoot as wet or moist surfaces freeze. Finally, it should be noted that even light winds can introduce wind chill, which causes a degree of coldness much more severe than the numerical temperature value would imply.[1]
– *Snow:* if there have been falls of snow before the persistent settled winter weather is established, the accumulated snow will not melt in the sub-freezing temperatures. Lying snow is always troublesome and any further falls will make the situation worse. If there is some daytime melting, the water/slush

1 Appendix B: Wind chill, see p. 220

then freezes to ice during the night, creating hazards for all who travel, whether in vehicle or on foot.

- *Fog and air quality:* settled conditions may also bring persistent and dense fog. This is more common in winter but may also occur in summer if the air is very moist. As explained above, the fog layer is trapped below the sinking air of a high pressure system. This mechanism also traps smoke and other pollution, seriously affecting air quality. Fog and poor air quality bring significant community dangers.
- *Persistent high temperatures:* on the other hand, if the season is summer and a high pressure situation is established, the largely cloudless conditions will boost daytime temperatures to very high levels. Such high temperatures can cause medical problems for those vulnerable to heat. Furthermore, a long period of this type of weather may well produce an extended drought. Drought may occur throughout the year but tends to be a more serious problem in summer when water demands are greater. Persistent drought can bring great inconvenience or hardship, even danger and death.

Consequentials of extreme weather: A focus on flooding

Of all the consequentials of severe weather events, flooding is the most common and destructive. The most severe floods cause total destruction, death and injury. In many cases, their effect is not only devastating but prolonged; in certain circumstances, the flooding can even be permanent.

The word 'flood' is one which brings immediate and disquieting images of misery because it usually refers to the ingress of water into an area designed to be dry. This is why the simple failure of a domestic water pipe brings such inconvenient and destructive problems, followed by that pervading dampness which seems to be so difficult to lose. However, these sort of domestic 'disasters' pale into insignificance when contrasted with seemingly all-too-frequent news images of deep and swirling flood waters, cruelly obliterating life and habitation. Such extreme events, truly 'disasters', bring death, injury and great hardship.

Why do floods happen? The first and most fundamental reason is that there is a great deal of liquid water on our planet. A glance at any world map or globe shows that there is much more water than land surface. In area terms, the distribution is 71% water surface, 29% land, so there is no shortage of the raw material for floods. Secondly, it is obvious that water is capable of movement; it does not stay exactly where it is shown on the map. Large volumes of water are able to move by a number of natural forces and this movement has the potential to cause flooding. There are two common forces that cause large-scale movement in the seas of the world:

- *Surface wind:* sea water moves as waves which are generated in the uppermost layer of the sea by wind blowing over the water surface. Some of the wind energy is transferred to the sea and waves are formed. The stronger the wind and the longer it blows, the higher the sea waves become. When the wind drops, the sea waves have acquired significant energy and so continue their movement, although they gradually flatten to become 'swell'; these are long, flat waves that can travel great distances. When there is a very strong wind blowing towards land, the sea waves are driven on to the coasts with great force. This may cause serious flooding, made all the more dangerous by the accompanying high winds.[1]
- *Tidal forces:* sea water also moves as tides that flow around the world. Strictly speaking, it is not the sea water that flows around the world; The water remains relatively still while the Earth turns on its 24 hour rotations. The effect of these movements is to produce a high tide at coastal areas every 12½ hours or so. The height of the tides varies during every 4-week period from the higher 'spring tides' to the rather lower 'neap tides'.

Tides are caused by the influence of the moon and the sun on the Earth's force of gravity. Their effect is to weaken the Earth's gravity and cause a huge heaping up of sea water to form a high

1 Appendix B: Sea waves, see p. 213

tide.[1] High tides, especially the highest spring tides, regularly cause coastal flooding. Tides also affect other large bodies of water not connected to the open sea but the tidal rise and fall is much smaller in such places and may be difficult to detect.

Tides alone rarely cause flooding. The variation between the neap and spring tides can be calculated for each area and defences constructed accordingly. However, when spring tides are augmented by another mechanism, like large onshore wind waves or a storm surge from far away, flooding can be serious.

– *Other flooding mechanisms:* there are other more extreme events that can cause catastrophic flooding over a wide area. Notable among these are tsunamis, which are very large and powerful sea waves produced by a strong earthquake under the sea bed. This was illustrated starkly on 26th December 2004 when a very strong undersea earthquake was recorded in the Indian Ocean off the Indonesian island of Sumatra. Tsunami waves up to 10m high were generated. The coastal areas of ten countries in the region were violently inundated. Over 250,000 people perished and many millions were made homeless.[2]

All the mechanisms described above can cause serious flooding although only one, wind waves, is actually a weather-generated event. The effects of these floods are limited to coastal areas except where land is very flat and low-lying. With very flat land, flooding can penetrate well inland and cover a considerable area. However, inland floods are almost always caused by another very common occurrence which affects all areas of the world to some degree or other. This very common occurrence is rain.

Put very simply, rain is liquid water that has been taken from the sea or other sources of water at ground level and then 'sprayed' back on to the surface as a significant feature of weather. Of course the 'rain' may fall in other forms, such as sleet, snow or hail. The meteorologist has a generic word for all of these – 'precipitation'.

---

1 Appendix B: Tides, see p. 214
2 Appendix B: Tsunamis, see p. 217

When rain is very heavy and prolonged, the sheer volume of water can cause local flooding in extreme cases; more usually, flooding happens because a large quantity of rainwater falls on high ground and quickly drains down to lower ground areas. The only other major source of inland flooding is the relatively sudden release of pent-up water, for example, the rapid melting of large volumes of snow or the failure of a dam, natural or man-made.[1]

Extreme weather in history

Weather in the Bible

There is a tremendous amount of weather in the Bible; previous chapters have presented examples of biblical texts which range across the whole weather spectrum; everything from the most intense storms of violent wind, torrential rain, tornados and thunderstorms to the finest, most balmy and nurturing conditions. However the most fascinating thing about biblical weather is its range of use. Weather is not used just for story telling or scene-setting but is also used to communicate complex Christian teachings, making them more accessible to the ordinary people of 2000 years ago. The inspired writers of the Bible achieved this by linking the most complex spiritual ideas to everyday weather images and concepts. It is a very effective strategy still working today.

A direct reference to weather makes its first appearance in the Bible in the second chapter of the book of Genesis with mention of the word 'rain'. Within the biblical story of the creation of the world, the text is: '… no shrub of the field had yet appeared on the earth and no plant of the field had yet sprung up, for the Lord God had not yet sent rain on the earth…' [2] This identified rain (water) as an essential requirement for plant life. Many of the Old Testament texts using weather events or imagery are intended to demonstrate the power of God.

---

1 Appendix B: Rainfall and flooding, see p. 212
2 Genesis 2:5

Weather features in many of the principal bible stories. The Flood, with its torrential and flooding rainfall is an obvious example. 'And rain fell on the earth for forty days and forty nights'. [1] The story emphasised how deep this destroying flood became: 'The waters rose and covered the mountains to a depth of more than 20 feet'.[2] The life story of Moses, a very important early Israelite leader, also contains weather imagery. When the Egyptian Pharaoh refused to release the enslaved Israelite people, God sent a series of devastating 'persuasions', many of which had links to weather. The most direct example sent violent hailstorms to destroy crops and fragile life: 'Throughout Egypt, hail struck everything in the fields – both men and animals; it beat down everything growing in the fields and stripped every tree'. [3] The rest of the Old Testament contains many other significant references to weather description, imagery and concepts.

In the New Testament, a number of the teaching stories of Jesus involved weather. Two of these have direct links, since they involve stormy winds affecting the Sea of Galilee; these winds whipped up waves that were dangerous to the small open fishing boats, a source of transport for Jesus and his disciples. In the first of these stories, Jesus was asleep in such a boat when he was awakened by terrified disciples in fear for their lives. Jesus rose and the text reported his words: '"Quiet! Be still!" Then the wind died down and it was completely calm'.[4] In the following verse, Jesus rebuked his disciples for their lack of faith.

Similarly, the very famous story of 'walking on the water' describes a similar situation of sudden storm. This time Jesus is not with the disciples in the boat but comes to help them by walking on the water. There was no dramatic command but when Jesus arrived the situation changed; 'Then he climbed into the boat with

---

1 Genesis 7:12
2 Genesis 7:20
3 Exodus 9:25
4 Mark 4:39

them and the wind died down.[1] Although the two stories use a similar weather device, the teachings are different. The first was to teach trust and faith; the second was to emphasise that Jesus will come to help when needed and restore peace and calmness to any situation.

Despite its status as an ancient document, there is real meteorology in the Bible. The book of Genesis has significant examples of careful weather observation. For instance the shepherd Jacob described dry semi-arid conditions when he claimed to be burnt up in the blazing heat of the day and so cold at night that he could not sleep. An accurate description of the weather in such areas.[2] There is also the realisation of cause and effect in weather. The Old Testament book of Zechariah provides a good example by linking showers of rain to 'bright clouds' – an accurate picture of the sunlit white puffy cumulus clouds from which showers fall.[3]

Finally, there are biblical weather forecasts! These are the words of Jesus: 'When evening comes, you say it will be fair weather, for the sky is red; and in the morning, you say today will be stormy, for the sky is red and overcast'.[4] That's right! It's the famous: 'Red skies at night, shepherd's delight; red skies in the morning, shepherd's warning', a weather saying still used by people today. This does not mean that Jesus had become a weather forecaster, as well as the saviour of the world; in fact the subsequent text makes it clear that he is quoting a well-known piece of weather lore, used by the people of biblical times. This was within a teaching about requests for miraculous 'signs'.

However, this piece of weather lore has a scientific basis. It can be correct for mid-latitude regions of the world. When the sun is low on the horizon and there are very high, wispy cirrus clouds in the sky, the refraction of sunlight from the ice crystals that form such clouds changes the clouds from white to red. Significantly,

---

1 Mark 6:51
2 Genesis 31:40
3 Zechariah 10:1
4 Matthew 16:2

these wispy cirrus clouds may herald the presence of a weather system.

The assumption of this piece of weather lore is that weather systems generally move from west to east. Therefore a rising sun in the east will light cirrus cloud in the west and may imply that a weather system is on its way. Conversely, a setting sun in the west will light cirrus cloud in the east which is being carried away from the region. This weather lore works, but only if the weather system is moving from the western quadrant to the eastern quadrant (which often it may not) and the cirrus cloud is actually the leading edge of a weather system – which again it may not be. Nevertheless, this is biblical weather forecasting!

Weather down the centuries

Before the advent of scientific weather observing, weather records were kept in narrative form in many places. Where people were involved in weather-sensitive activities, quite detailed records of weather, especially bad weather, would be maintained. For instance, every sea-going ship would include weather records in the ship's log. Ports all over the world kept similar records. If severe weather affected any place, a detailed record of the events would be kept within its local history. However there was little general publication of this information.

It is obvious from these rather sparse historical records that severe weather has happened down the ages. However many historical global records refer only to the most severe events. These are disasters which are not associated with extreme weather but with the serious devastation caused by earthquakes, volcanoes and tsunamis. Such events are reported as total catastrophes, frequently accompanied by huge death tolls.

However there are early records of some weather-associated disasters. An extreme example of shipwreck occurred when the Spanish Armada, having been engaged in battle with the English Navy in 1588, was then struck by one or more severe storms on its return voyage to Spain. It is recorded that more than 50 ships were sunk with the loss of 20,000 lives.

Tropical storm damage with very large loss of life was reported several times in the 19th century in regions of India and China, where at least half a million lives were lost. Worse still, a further million people were killed in China when the Yellow River flooded severely. On the other side of the coin, there are also several 19th century reports of devastating drought in other parts of the world where large loss of life was reported.

Recent records

Here are details of some notable flood events recorded in more recent times:

- *The Great USA Flood of 1993:* the summer months of 1993 brought much flooding to mid-western areas of the USA, when the upper streams and tributaries of the Mississippi and the Missouri rivers burst their banks. As many as 150 streams and rivers were involved in this flood situation. Over 1 million sq.km. of land was flooded and there was a loss of 50 lives with many more injured. Large areas of agricultural land were lost and thousands of homes destroyed. The floods persisted for six months in some locations. The damage totalled over 15 million US dollars. This was judged to be the most serious flood ever to affect the USA.

  The meteorology indicated flooding was inevitable. The weather observations from that time in 1993 report extensive precipitation in the region, not only in the summer months but over the previous months too. There are reports of high water tables, saturated ground and filled reservoirs, lakes and streams. Then, in the flood period itself, the area was subjected to a series of very powerful storm systems, each producing violent precipitation that fell on saturated ground, making serious flooding inevitable.
- *Severe flooding in Europe, August 2002:* very active weather systems that had originated over the Atlantic Sea moved east to affect large areas of western and central Europe and also some of the Mediterranean countries. Intense storms were generated, with persistent, torrential rain. There was severe flooding in parts of Germany, Austria, Romania and the Czech Republic. A further weather system development affected areas further south, notably in Spain and Italy.

The hydrological reports across this region recorded high water tables and saturated ground, making flooding inevitable when more heavy rain occurred. The situation in many areas was made worse by flash flooding when the rainfall became most intense. Flash flooding occurs when large quantities of rainwater are channelled down steep river courses to valley or plains areas below. During this month in 2002, many rivers burst their banks and there was considerable devastation throughout the regions. Over 100 fatalities were recorded.

- *Flooding in other regions:* periodically, there have been reports of flooding in many other mid-latitude regions (in Northern Asia, South America, Southern Africa, Australia and New Zealand). Unless catastrophic, these reports tend to be brief. They usually describe a situation of river flooding caused by extended periods of torrential rain.

Of course flooding is extremely common in monsoon regions. These are the seasons of torrential rain that occur in tropical and sub-tropical latitudes where there is a vast amount of heat input and plenty atmospheric moisture available. The word 'monsoon' actually refers to the wind flows which are generated in these seasons and, originally, referred only to the Indian Ocean region. Nowadays, the word is also used to describe rainy seasons in most other parts of the tropical and sub-tropical zones. When severe flooding occurs, the poverty in such regions is often a significant factor in the loss of life and devastation which occurs, because the systems and structures which could protect the population are inadequate or largely non-existent.

Past extreme values of weather elements

Over many years, records of extreme weather events have been compiled, covering various weather elements like temperature, rainfall amount, and wind speed. This data is available from many sources, including reference books and websites.

- *Surface air temperature*: the highest recorded temperature continues to be listed in most sources as El Azzizia in Libya, North Africa, with a value of 58 DegC (136 DegF), recorded in 1922. El Azzizia is located in the northern part of the Sahara Desert.

Death Valley, California, USA is not far behind El Azzizia with 57 DegC (134 DegF) recorded in 1913. Death Valley is located in the Mojave Desert of the USA and the area is notable for being 85.5m (282ft) below sea level.

The lowest temperature record is held by Vostok in Antarctica with a temperature minimum of -89.2 DegC (-128.6 DegF) on 21st July 1983. It is relevant to note that this station was commissioned in 1957, so there are no weather records before then.

– *Rainfall amount and intensity*: the records indicate that the highest annual rainfall total was 26,461mm (1042in) at Cherrapunji, India, back in the 19th century, in 1860-61. Cherrapunji is set in the hill regions of northeast India, at a height of 1484m (4869ft) above sea level.

  Foc-Foc, La Reunion Island (an Indian Ocean island east of Madagascar) experienced the most rainfall in 24 hours - 1825mm (71.9in) fell during 7-8 January 1966.

  The most intense rain was recorded in Unionville, Maryland, USA when 3.1mm (1.23in) fell in one minute on 4th July, 1956.

– *Wind speed:* Thule in Greenland holds the surface wind record for a low elevation weather station. On the 8thMarch, 1972, 333 kph (208 mph) was recorded there. (Records began at Thule in 1961.)

  At higher ground levels, the wind speed record is held by Mount Washington, New Hampshire, USA, whose peak rises to 1917m (6288ft) above sea level. This is the highest mountain in the northeastern part of the USA. A wind speed of 370 kph (231 mph) was recorded there in 1934.

So, should we expect more 'extreme weather events'?

In a 2001 report, the IPCC stated 'An increasing body of observations gives a collective picture of a warming world and other changes in the climate system.' Their latest report (2009) strengthens and refines that view, stating: 'Warming of the climate system is unequivocal, as is now evident from observations of increases in global average air and ocean temperatures, widespread melting of snow and ice and a rising global average sea level.'

Regarding their prediction of increasing temperature, this is now described as 'virtually certain' while the frequency and intensity of severe weather ('more extreme weather events') is predicted as 'very likely'.

So it should be noted that the IPCC warning of 'more extreme weather' is not unequivocal, even although it has been expressed with increasing confidence. However, if the increasing frequency and intensity of severe weather events is linked simply and directly with the progress of global warming, then we would expect to see most of the 19th and 20th century extreme weather 'records' broken by new 'records' set in more recent years.

However that expectation does not appear to be met. The historical record of extremes presented earlier in this chapter does not offer evidence of increasingly severe weather over time; in fact most of the extreme 'record' values have been shown to occur in the early or mid-20th century. On the other hand, recent meteorological reviews assess the year 2006 as globally the hottest year on record, with the summers of western Europe and Australia singled out as 'extremely hot'.

So, at present (2010), the answer to the question: 'Will there be more extreme weather events?' cannot be an unequivocal 'Yes' but rather 'It seems increasingly likely'.

6

Could We Control The Weather?

Weather control is a subject with a long history. People have wanted to control the weather for many thousands of years – that is, change it to suit their own purposes and desires. Humanity has tried many ways in the past; there have been many earnest appeals to deities, sacrifices offered, special rituals, rain dances (or stop-the-rain dances!), chants, prayers and incantations. All facets of the weather have been involved. As well as the desire for rain or lack of it, there have been earnest requests for wind or calm, sunshine or cloudiness, heat or cold, dryness or humidity, etc. All this has happened throughout history and indeed still happens today to some degree.

The opportunity to be rather more proactive came in the earlier part of the last century when there was a growing scientific understanding of weather. The late 19th century had been a time of exploration into the physics of weather but it took several more decades before fundamental weather processes were understood in a reasonably coordinated way. By this time, it was recognised that large weather systems contain vast amounts of energy. To change these systems in any way implied the input of similar quantities of energy and this was known to be an impossibility. Even relatively small weather events like individual rain showers involve a great deal of energy.

On the other hand, studies into the formation of precipitation suggested there might be a way to influence natural rainfall. Water is a fundamental of life and the delivery or prevention of rain at a particular time and place was recognised as a way to help areas that were afflicted by too much or too little rainfall. Certainly, in harsh and extreme weather environments, rainfall control of this type would be of great benefit to the life of the population in general. All these considerations led to the concept, theory and practice of a system of possible rainfall control. This came to be known as 'cloud seeding'.

Rainfall control by 'cloud seeding'

Cloud seeding was conceived in the USA in the 1930s. In the following decades, experiments were pursued with enthusiasm (not only in the USA) but results were far from conclusive. In those days, military interest provided a significant impetus for research. It was thought that weather control could give great battlefield advantage. For instance the enemy could be bogged down in continuous torrential rain and greatly weakened; or perhaps subjected to drought conditions that would generate huge logistical problems to meet their army's water requirements.

It is a fact that research projects seen to have military advantages are often resourced generously and progressed with great vigour. However since 1977, the UN has outlawed the use of weather modification as a military weapon. Since then, the focus of all such work has been largely on agriculture or general population support.

What is 'cloud seeding'?

Cloud seeding is a procedure which attempts to change the natural rainfall that may happen in a particular weather event. In theory, the procedure may be applied to a range of rainfall situations; it may be applied to everything from small individual shower clouds to the cloud formations of larger weather systems. Many people associate cloud seeding only with boosting rainfall but it may also be used to suppress rainfall. In the latter case, the rain is 'encouraged' to fall before it arrives over the target area. Also, the same theory of cloud seeding may be applied in an attempt to change the character of the precipitation, for instance to prevent hail from falling on a particular area. There are some places in the world where violent hailstones are a frequent occurrence which causes a great deal of agricultural crop damage.

Of course, to increase the rainfall of a very dry arid area of the world would be of great benefit to the local population. For some people who live with frequent drought conditions, the daily task of acquiring sufficient water is a major part of their lives and a great drain on their personal energy resources.

How is cloud seeding done?

There are two basic types of cloud seeding and the choice of which type to use is dependent upon the temperature structure of the clouds to be treated. The temperature structure of a cloud allows it to be categorised either as a 'warm' cloud or a 'cold' cloud. Because air pressure decreases with height, air expands as it rises further from the surface of the Earth. Such expansion requires energy and that energy is provided from the heat in the air, which cools as a result.

When cloud forms in the atmosphere, it also becomes colder with height. If the cooling within the cloud never takes the temperature below freezing point (Zero DegC), then that cloud is categorised as a 'warm' cloud. This type of cloud is commonplace in the Tropical and Sub-tropical regions of the world but may also occur in mid-latitude regions during the summer season. If however the cooling within the cloud takes the temperature of its upper levels below freezing point, then the cloud becomes a 'cold' cloud. This categorisation is not only important for cloud seeding but for meteorology in general because the precipitation formation mechanisms within warm and cold clouds are significantly different.

If the cloud is a 'warm' cloud (all its water droplets are above freezing-point), cloud seeding consists of spraying hygroscopic particles into it; these are particles that absorb water, such as salt. In all clouds, condensation can only occur if there is a sufficient supply of 'condensation nuclei' because saturated water vapour needs something to condense upon. That 'something' may be sea salt, dust, sand or carbon particles, all of which have been drawn into the cloud by air rising from the Earth's surface. In many clouds, there is a shortfall of condensation nuclei and some saturated water vapour is unable to condense. In meteorology, this water vapour is referred to as being 'supersaturated'.

So the idea behind this type of cloud seeding is to provide extra condensation nuclei material so that all the saturated and supersaturated water vapour will be able to condense and form water droplets (clouds). Rain occurs when the water droplets collide and merge to form bigger water drops. When these large water drops become too heavy to be supported by the upcurrents in the cloud, they fall out as rain. The theory is that the seeded cloud will be changed to contain more and bigger water droplets, so the rain that falls from the cloud will be heavier and last longer.

However many clouds are not 'warm' clouds because they extend above the freezing level (the height where the temperature drops below zero DegC) and are defined as 'cold' clouds. Above the freezing level, cold clouds are formed partly of ice crystals but there will also be some water droplets that have stayed in the liquid water state despite being below freezing point. These water droplets are defined as 'supercooled'; this means that they are ready to freeze but they cannot do so until they encountered suitable particles around which to freeze. What they require are particles known as 'freezing nuclei'.

When the supercooled water droplets encounter already-formed ice crystals, they immediately freeze upon them, making these ice crystals bigger. When the ice crystals are too heavy to be supported by the upcurrents in the cloud, they fall out as precipitation, usually (but not always) melting to water drops as they fall into the warmer air nearer the surface of the Earth. This is by far the most common natural rain-making process.

The cloud seeding procedure for cold clouds is to spray particles into the cloud to act as an additional supply of freezing nuclei. The particles are designed to mimic the structure of ice. Common materials used for this purpose are solid carbon dioxide (dry ice) or silver iodide. At the appropriate time and place within the cloud, the introduction of these particles should in theory result in more freezing, bigger ice crystals and heavier precipitation, delivering in due course a greater amount of rain to the ground below.

The usual method of both types of cloud seeding is by aircraft. At the appropriate location, the seeding material is either released from the aircraft in a stream or it may be fired as a particle-filled explosive shell that is detonated shortly after. An alternative method is to fire the explosive shell from the ground and detonate it at an appropriate height. This method is obviously less flexible when compared to aircraft delivery but both methods have been used in the past. Obviously, the seeding material has to be placed within the cloud at the right place for there to be any chance of success.

Does it work?

Seeding warm and cold clouds has the potential to work successfully in certain highly specific circumstances. However many long-term tests have not indicated success with any sort of reliability and many experiments have been judged a failure. Some scientists have pointed out that the atmosphere is far too complex a structure to respond reliably to such simplistic hit-and-miss operations.

Today, there are still a number of nations who experiment with cloud seeding. Israel has claimed success with a 15% rainfall enhancement in a dry area of its country; Mexico continues with experiments and also claims some success. Texas, in the USA, has conducted many experiments over the years; the results are variable and disputed routinely.

Research is still carried out to some extent in other countries, mainly those who have large areas of land that suffer from a lack of rain. The specialist companies who carry out seeding operations are still awarded contracts by other states and authorities in various parts of the world. For obvious reasons, there tend to be claims of success from these commercial sources. Certainly, any claim of success based on a single or small number of events is invalid. Most people would be highly suspicious of a single sequence of 'rain dance performed - rain arrived - total success achieved' The words 'cloud seeding' could be substituted for 'rain dance'!

So the extent of the influence of cloud seeding is very limited. Seeding done within a particular cloud can only affect that cloud. A series of seeding operations may affect a system of clouds but the chance of each cloud being poised to react similarly is remote. This is why cloud seeding has had so little positive effect over its long history.

Are there other ways to change the weather deliberately?

It is tempting to say 'no' because the amount of energy needed to alter any other element of weather is so enormous. However there is one way that has been attempted down the ages. This does not change the weather in the sense that cloud seeding tries to do but it attempts to change the weather conditions at a particular local area for a limited period of time. It is the deployment and continuing generation of a smoke screen; this has often been used to achieve military advantage.

Obviously, thick smoke will have a great effect upon those subjected to it. They will not be able to see their adversaries and this will put them in great danger. It may affect the operation of their equipment and cause it and them to become technologically impaired. Perhaps more importantly, it may affect their health. It may incapacitate or even kill. However the smoke screen can only be maintained if the general weather permits. A changing wind would blow the smoke away – or even blow it back on those who are generating it. A calm wind will result in the smoke particles sinking to the ground.

Even so, it has to be admitted that a large and persistent smoke screen does indeed alter the local weather to some limited degree. This has been seen in the deliberate or accidental burning of oil wells, for instance; the ignited oil provides a continuous supply of dense, choking and obscuring smoke that spreads out to affect very large areas of ground downwind. Large forest fires have the same effect and their deliberate firing is not unknown. Probably the largest scale obscuration comes from volcanic action and this can certainly affect the weather over a large area.

Aside from the direct effects already discussed, the general weather within a smoke polluted zone is likely to be colder because the sun is obscured. Technically, therefore, the deliberate generation of a dense and persistent smoke screen is a method of changing the weather to some degree, although its effect is very limited and highly unreliable.

Weather control in the Bible

God and the weather

The Bible leaves us in no doubt that God is in charge of the weather and able to control it. There are many occasions in the Bible when God is reported to use elements of weather to achieve particular purposes. This ranges from huge, violent storms sent to bring retribution and punishment through to his gentle and loving delivery of life-giving rain or balmy sunshine to his favoured people; there is also his personal use of cloud as an obscuring agent during the times he wishes to communicate directly with people.

- *Wind, rain, storms and more...* the Old Testament book of 1 Samuel provides a typical example: The Israelites were disobedient to God and Samuel (an important early prophet) was determined to teach them a lesson and bring them to their senses. At the time of wheat harvest, when the weather is normally dry and settled, Samuel gathered the people together and then asked God to send 'thunder and rain' upon them. This was delivered immediately. Unsurprisingly, the people were convinced and returned to worship God - and of course Samuel's reputation was greatly enhanced, too.[1]

---

1 1 Samuel 12:16-18

An even more extreme example is given in the book of Ezekiel: Gog (Prince of the land of Magog) had attacked Israel with a huge army. God defended his people by waging war upon the attackers. He sent extreme elements of weather to afflict them, grievous storms including wind, rain, hail, lightning and thunder; in addition there were earthquakes, plagues and other illnesses. As a result, Gog and his army were defeated and Israel was saved from destruction.[1]

- *Drought:* God also used his control of weather to produce drought conditions in a number of places in the Old Testament. The book of Haggai records how the people of Israel returned to Judah after their Babylonian exile. They were instructed by God to rebuild the Temple in Jerusalem. However the Israelites ignored this instruction in favour or building their own houses and farms. God then 'encouraged' them to start work on the Temple by imposing a drought upon them; this, of course, caused many problems including the loss of all their crops. The Bible records that the people decided to start work on the Temple immediately! [2]
- *Cloud:* God is first reported to use cloud as an obscuring agent in the book of Exodus when he was guiding the Israelites on their journey from Egypt to the 'Promised Land'. In this case, God was obscured by a 'pillar of cloud' during the day which became a 'pillar of fire' at night.[3] The 'obscuring cloud' device is used many more times in Exodus. God descended to Mount Sinai in a cloud when Moses received the Ten Commandments.[4] God concealed in clouds was also referred to in the Psalms and in Nehemiah.[5]

There is similar use of cloud in the New Testament. For instance, Matthew's gospel tells of the events of the Transfiguration of Jesus, when he was affirmed as God's son. Jesus had taken three

---

1 Ezekiel 38:17-23
2 Haggai 1:7-11
3 Exodus 13:21
4 Exodus 19:16
5 Psalm 78:14, Nehemiah 9:12

of his disciples to the top of a high mountain. The text reads: '…a bright cloud enveloped them and a voice from the cloud said: "This is my Son, who I love; with him I am well pleased. Listen to him!"' [1]

Jesus and weather

- *Calming the storm*: there are two occasions recorded in the New Testament when Jesus demonstrated his control of weather. The first is known as the 'Calming of the Storm.' Jesus and his disciples were crossing the Sea of Galilee in a small open boat. Jesus was asleep when a sudden squall brought very strong winds which soon whipped up the sea into large waves. The disciples struggled to control the small boat and became increasingly afraid that it would sink. They woke Jesus in great fear. The culmination of the story is his dramatic command to the wind and waves: 'Be quiet, be still!' There was instant obedience.[2]

  The other occasion is during the very well-known 'Walking on the Water' account. Here, Jesus was not with the disciples in the boat. A very strong wind started to blow and the waves were rising dangerously. The disciples were again struggling with the small boat and were in great fear. Then Jesus came to them, walking on the water. When he climbed into the boat, the wind and waves were stilled. This time, there was no command spoken to the wind and waves but his presence in the boat imposed an immediate calm.[3]
- *Use of cloud:* in the New Testament, Jesus referred to the return of the 'Son of Man' – a term he often used to describe himself: '… they will see the Son of Man coming in (or on) a cloud in power and great glory.'[4] (However, when this is mentioned in the book of Revelation, the text is: '…and seated on the cloud was one like a son of man…' [5]). In Revelation, there

---

1 Matthew 17:5
2 Mark 4:35-41
3 Mark 6:45-51
4 Luke 21:27
5 Revelation 14:14

is a clear description of a man's figure seated ON (upon) the cloud. While the Greek word used in the Matthew text may be translated in English as 'in' or 'on', the Greek word used in Revelation means 'on' or 'upon' only.

In fact this is an example of a very clever use of imagery description. Throughout the Bible, God is always described as being concealed in (within) the cloud. Examples of this were given in the section 'God and the weather' above. This device is always employed because it is impossible for anyone to picture God. However, Jesus can be pictured as a man, even if we cannot know exactly what he looked like. In this sort of image, we can still present him as a visible human figure sitting upon the clouds. This is a fine example of the communication of spiritual ideas by the use of weather imagery.

Broad-scale weather control today: changing the climate

The relationship between weather and climate

There is a clear difference between 'weather' and 'climate'. Both these words appear in the IPCC warnings and there is a need to recognise the difference. 'Weather' may be defined as the atmospheric conditions that exist in the open air. Although this will commonly refer to the conditions at or near the ground, it can also be the weather conditions in the atmosphere at considerable height. Both commercial and military aircraft are interested in the weather conditions that will affect them at their cruising levels.

Basically, 'weather' is a combination of the conditions which we find anytime we go outside into the free air. People have many terms to describe the weather they find, using words such as 'cold, windy, cloudy, raining, humid, etc.'. Scientific meteorology defines weather conditions by making observations that are recorded in various scientific scales; temperature is recorded in Degrees Celsius or Degrees Fahrenheit (DegC, DegF), wind speed in knots (kt) or metres per second (m/s), rainfall in inches (in) or millimetres (mm), etc. However assessed or measured, the result is an observation of the 'weather' at that place and at that time.

'Climate' is the general weather experienced by an area or region. In most regions of the world, there are specific seasons of weather (summer, winter, etc.). People define their climate by their community experience over an extended period. However, climate is highly related to weather, since it is the accumulation of individual weather events over a long period of time.

People describe climate in the same terms as weather, although there are some additional descriptions like 'temperate', 'maritime' and 'continental'. Scientific meteorology defines climate by averaging its weather observation measurements; the climate values of one year are often contrasted with long-term averages of climate to determine whether the year was 'cold', 'average' or 'wet', etc.

So climate and weather are very closely related. Climate is a perception of weather that can be constructed from a record of weather events. Climate is a product of weather; it does not exist without it.[1]

Imposing a different climate

Theoretically, there are three ways to change the climate of an area. The first is to change individual weather events. Done consistently, these changed events would cumulate over an extended period to impose a different climate. Earlier in this chapter, the discussion of cloud seeding showed that the alteration of rainfall events is difficult, unreliable and likely to be unsuccessful. Apart from the use of smoke (which is even more unreliable than cloud seeding), there are no other ways of changing individual weather events.

The second way is to deliberately alter the vegetation cover of the land, although this will need to be done over an extensive area if it is to be effective in a climate sense. Changing to a rich cover of growing plants – everything from thick grassland to dense forest – will cause heat and moisture to be added to the atmosphere, especially if irrigation is installed. The heating and moistening of the

---

1 Appendix B: Climate and weather, see p.209

air will generate uplift, cloud and precipitation to some degree and this will contribute to the determination of the climate in that area. Alternatively, removing vegetation (deforestation, land clearing for agriculture) will have the opposite effect, with cloud and rainfall reduced. Severe land clearance will impose a barren climate upon the land, even desertification.

The third way would be to alter the broad-scale weather patterns from previous norms. This would alter the behaviour of weather systems and change the climate. The formation and behaviour of broad-scale weather patterns is very complex. Such patterns cover huge areas of the globe and involve a great amount of energy. Mankind has never *deliberately* affected such weather patterns but current warnings of climate change show that inadvertent changes are possible. This will be discussed in more detail in Chapter 7.

Achieving reversion to a previous climate

In a real sense, this is the ultimate weather control issue which is assuming greater importance as the current situation becomes clearer. It is already apparent that global warming has substantially altered the sea ice distributions of polar and high latitude regions. This, and the general rise in sea temperature, has caused very low-lying areas across the world to be flooded; in addition, the apparently small land and sea temperature changes have already affected plant and animal life significantly.

Reversion of these effects could only be achieved by the restoration of colder sea temperatures and the reformation of ice at the Poles and elsewhere. Total reversion seems unlikely in the short term but a slowing of global warming should retard further changes. Current efforts concentrate upon slowing the global warming trend, mainly by concentrating upon reducing carbon emissions.

Global warming certainly has the potential to affect and alter weather patterns but this is a much more difficult matter to assess. If climates were altered in this way, it is certain that reversion to previous weather or climate patterns would be a most difficult, even impossible task.

God's people and weather control

Does God want his people to be able to control the weather or the climate of an area? The Bible may suggest an answer. The Old Testament is full of accounts of the people's disobedience to God's will and how he becomes angry and sometimes despairing when this happens. The people are guilty of a whole range of truly negative actions (sins) including selfishness, dishonesty, cruelty, weakness and many others.

However, the Bible establishes the worst sin of humanity early in the book of Genesis, namely any attempt to achieve equality with God. Essentially, this is what happened in the symbolism of the Garden of Eden when the forbidden fruit was eaten. Perhaps this is even more clearly illustrated by the Tower of Babel story,[1] when the people were not only being disobedient to the will of God but had returned to the sinful ambition of achieving equality with him.

In the story, God had noted that his people were multiplying satisfactorily and now he wanted them to scatter and occupy the whole Earth. Unfortunately, this was not what the people wanted! They all shared the same language and culture; they were leading comfortable lives and were well-settled in a pleasant land. So they decided to ignore God and live together in one large magnificent city which they set about building.

Then they made the situation much worse. They decided to build a very high tower in the centre of their city, a huge tower which would 'reach to the heavens', thus putting them on a level with God and achieving the equality they sought. Since God would then have no authority over them, they could do what they liked – and this was their goal. In particular, it would mean that they could all stay together forever in their pleasant lifestyle. The story tells how God observed what was happening and took swift and decisive action. He confused their single language so that they

---

1 Genesis 11:1-8

could no longer communicate with each other and then physically scattered them all over the Earth.

So the symbolism of the Tower of Babel suggests that God will not permit humankind to 'go their own way' to the ultimate extent. On the other hand, we know that God created mankind to have free will. Fortunately, it is possible to reconcile these two statements. Perhaps the analogy of a small child and his or her parents will illustrate this reconciliation.

The parents want the child to develop optimally; for this to happen, he or she must have freedom (free will) to discover and develop. But the free will cannot extend into actions of danger which have the potential to injure or kill, e.g. playing with a sharp knife. If the parents see that the child's actions lead to such dangers, they will prevent that particular action. Although this will frustrate the child, it is clearly an action of love and nurture.

Similarly, God wishes his people to have free will and to be intelligent and inventive; indeed, this is the way the world advances and develops. By allowing free will, God demonstrates his great love and nurture for his people. However he will not allow actions which pose extreme danger for them. If people actually achieved 'equality' with God, the father and child relationship would be destroyed and separation would occur. God knows that the loss of his love and nurture would undoubtedly be a very great danger for his people (children). In love, he is not willing to allow this to happen.

So God is willing to allow his people to acquire the scientific knowledge needed to protect themselves and the world. In many ways, this is already happening with global warming, weather and climate change; there has already been a great deal of study and inventiveness applied to the analyses and solutions of these particular problems. Perhaps even more importantly, it is increasingly appreciated that problems which are truly global require truly global solutions. This in turn needs truly global cooperation between the nations of the world. It also means personal engagement, support and compassion for others, in a global sense. All encapsulated in three words: 'love one another....'

7

How Does It All Add Up?

Global warming, sea currents and climate change

'Catastrophic climate change' was mentioned in Chapter 1 as one element of the IPCC warnings. Although it is a valid part of the possible effects of global warming, the danger is currently judged to be 'unlikely' for most of the 21st century. Catastrophic climate change is bound up with the vast expanses of the world's oceans, in particular with the effects of long-established sea currents upon weather and climate.

It has long been known that the world climate is greatly affected by the huge currents in the oceans. Any alteration to these sea currents will impose climate changes. For instance, a region affected by warm sea currents (e.g. the warm Gulf Stream of the Atlantic Ocean) would turn much colder if the warm currents were cut off or weakened. The important point here is that the change from one climate to another would be quite sudden. Furthermore, because a very large and conservative feature had been altered, it would be extremely difficult to reverse, perhaps even impossible.

Sea current organisation:

The oceans are vast energy stores and the major sea currents such as the Gulf Stream and the North Pacific Current have long been known as a major means of energy transfer around the world. It will be recalled that 71% of the Earth's surface is covered with water. These vast sea currents have a major effect on the weather and climate of the regions around them and, because they also impinge upon atmospheric weather patterns, they extend their influence around the world to areas which are far from sea shores.

Sea currents are a response to temperature changes across and within the oceans and also to changing levels of salinity. The key value is density; cold water is more dense than warm water and salt water is more dense than fresh water. Variations in density generate flow movements in the sea water. The movements then cumulate to become major currents within huge areas of the oceans. In fact they organise themselves into vast circulations.

For instance the warm current of the Gulf Stream flows north-east across the Atlantic at relatively shallow levels, sinks down in the cold waters of the Arctic and then flows back south in the depths of the ocean. The salinity of the ocean water plays an essential part in this circulation and the cooling effects of the polar ice fields are also of great importance. However it is known that the world's sea currents are even more organised than this simple circulation would imply.

It has been shown that the world's major sea currents are in fact linked together to form one vast circulation that flows through all the major oceans. In this vast global circulation, warm sea water flows in the upper layers of the sea to return as cold currents deep in the oceans near the sea bed. This immense circulation is known as the 'Great Ocean Conveyor Belt' and its vast flows pass through the oceans of the world, traversing both southern and northern hemispheres. Just like the atmospheric airflows which control weather developments, the flow strength and specific location of the Great Ocean Conveyor Belt varies within the ocean volumes.

However the greater density of water ensures that these sea current variations are more conservative than the high level atmospheric airflows which are in a state of constant and rapid change.

Effect of global warming on sea currents

Global warming has raised sea temperatures. Measurements have revealed that this warming has penetrated deep into the oceans, with the top 1000ft of the sea water depth showing a temperature increase of 0.3 DegC on average during the last century. Northern polar regions have been affected even more, with the polar ice cap melting significantly. The melting ice has added huge volumes of fresh water into the Arctic seas and reduced their salinity levels; this changes the key value of density and affects sea current movement.

More recently, the situation has been made worse by changes in surface reflectivity in polar regions. As the highly reflective sea ice surfaces are replaced by large areas of open water, more solar heat is being absorbed into the water, accelerating the warming process further. Furthermore, it is known that the warming of the oceans reduces their potential to absorb carbon dioxide; the oceans have a vertical current mechanism to remove recently absorbed carbon dioxide from surface layers and store it in much deeper waters. The surface layers are then available for more absorption. Rising ocean temperatures inhibits this essential process in the sea and the result is less carbon dioxide absorption.

The combination of these changes of temperature and salinity has the potential to radically alter the normal patterns of the Great Ocean Conveyor Belt. If the sea current alterations are greater than the normal variations that occur and if they are sufficiently sustained, a sudden change in normal climate patterns could occur for some regions of the world. The basis of this prediction is the advice of complex computer models that have been run with various assumptions of global warming values.

In particular, it is thought that the Atlantic Gulf Stream is vulnerable to such changes; the weakening or total disruption of the Gulf Stream would result in a much colder climate being imposed on north-western Europe. However, the latest IPCC warning suggests that this sort of catastrophic climate change is 'unlikely' for much of the 21st century. It is, however, a factor that needs to be watched carefully.

Climate oscillations

Another strand in the discussion of climate change recognises that there are oscillations in climate patterns in various parts of the world. It is stressed that these oscillations are normal events that have been recognised for many hundreds of years. The oscillations manifest themselves as disruptions to normal ocean/atmospheric conditions and these can cause weather pattern reversals. The most marked of these features occur in the Tropics and these have been studied for many decades. More recently, oscillations have been noted in other parts of the world but information on these is much more limited.

– *ENSO [El Niño and the Southern Oscillation]*: of all the identified oscillations, *El Niño* (the Spanish words mean 'the boy-child') has emerged as the best known and arguably most significant factor in current weather and climate change considerations. *El Niño* is described traditionally as an unusual sea current of very warm surface water which periodically affects the area around western Peru. This causes a period of very wet weather, often with destructive coastal and mountain flooding. In the past, *El Niño* tended to happen every 5 years or so, starting in the month of December and usually persisting for several months. For most of the last century, *El Niño* was regarded as a local weather aberration that was of significance only for that part of South America and its environs.

  For some time now, however, the *El Niño* phenomenon has been recognised as a more significant and important climate feature whose effect is much more extensive than previously thought. The *El Niño* anomaly is now known to affect much of the Pacific Ocean and extend into areas beyond; today, it is more accurately described as a major disruption in the ocean/atmosphere structure of the tropical Pacific region.

  The 'normal' weather pattern across the tropical Pacific Ocean establishes easterly surface winds which carry warm water to the western Pacific region; this in turn generates mainly hot, unsettled, wet weather for that area. By contrast, the surface waters in the eastern Pacific remain cool; here, the normal weather conditions stay largely dry and settled. Periodically, the wind flow in the tropical Pacific reverses and the *El Niño* phenomenon is generated. This imposes the weather reversal known as the Southern Oscillation, with the western and eastern Pacific regions effectively exchanging normal weather types.

So the marked *El Niño* change in Pacific sea temperatures alters the weather patterns over a very large region, bringing not only the expected storms and floods to coastal South America but uncharacteristically dry conditions to large areas of the western Pacific. Drought conditions are produced over a huge area, with devastating effects on normal agriculture. Notably, there are occasions of very damaging brush fires in Australia; these are now linked directly to *El Niño* events.

In addition, a further complication may occur. At times, the disruption to the normal Pacific conditions may act to make Pacific waters considerably colder than usual. When this happens, another phenomenon has been recognised. This has been called *La Niña* (Spanish, 'the girl-child'). In general, the effect of *La Niña* is opposite to that of *El Niño* and it is a much more persistent feature than *El Niño*. Although the *La Niña* situation implies a return to 'normal' weather patterns, in fact it produces a more extreme version of normality; dry periods become drier, wet events become much wetter.

Significantly, scientists who study the *El Niño/La Niña* phenomena report that these disruptions are now happening with much greater frequency and, when they occur, the effects are much more powerful and long-lasting; a link with global warming is claimed. It can be seen that the *El Niño* phenomenon acts to change the weather over a very large area of the globe. More persistent *El Niño* effects would certainly change the climate of the regions defined; some say they may have done so already and that the influence of the phenomenon may well become more powerful and spread to even more parts of the world.

– *Other Oscillations*: several other weather oscillations have been identified. There is an Arctic Oscillation which affects pressure conditions in the more northerly part of the northern hemisphere; this phenomenon is linked to oscillations in the North Atlantic and North Pacific. Further oscillations have been suggested for the Pacific, the landmass of North America and in other parts of the Tropics. Of course, alterations to the normal sequence of these oscillations would in time have some effect on the climate of the world.

Are there suggestions of catastrophic climate change in the Bible?

There are many occasions when God used elements of weather to punish people for their transgressions. These actions were sometimes against his own people but at other times were used against the enemies of his people - a dramatic sending in of the 'weather' cavalry! Usually these can be defined as 'extreme weather events'.

On the other hand, the story of the Flood[1] is an example of climate change with a vengeance. Here, the climate of a pleasant, productive land was altered catastrophically by huge amounts of rain (although other flooding mechanisms are also mentioned in the Bible story). The very dramatic account then tells how all this 'rain' flooded the land to very great depth, and, apart from the occupants of Noah's Ark, all life on Earth was wiped out completely.

The Bible does not specify the geographical extent or the exact location of the Flood and, in theological/scientific discussions, it has been attributed to various large geographical areas or proposed as a 'whole world' event. Alternatively, it may be a very dramatic story that uses weather imagery to illustrate and communicate one of God's extreme mysteries, involving the destruction and re-creation of life. This has been fully investigated in a previous book.[2]

---

1 Genesis 6-9: all verses

2 The author has explored all aspects of the Flood in his book 'Divine Weather' published by Highland Books

The imposition of drought is another example of climate change. When little or no rain comes at the time of the 'rainy' season, this is a disaster initially for crops and subsequently for animal life. The failure of crops means severe food shortages; insufficient water for farm animals means that they are likely to perish. In the most extended droughts, people may die of thirst or be struck down by disease brought on by their weakened state. Meteorology knows that conditions of drought in a normal rainy season will be the result of a significant change in weather patterns, with the normal rain-producing systems weakened or transferred elsewhere by altered broad-scale flow patterns high in the atmosphere.

In the Bible, God uses the weapon of drought a number of times, although some of the texts are warnings. Such a warning is given to the Jews when God specifies they are to go to Jerusalem to worship him. Failure to do so will result in drought.[1] By contrast, on other occasions the disobedient people of God were subjected to actual drought so that their harvest would be destroyed. [2]

Adding in all the other pollution problems

In addition to global warming, three other pollution problems were identified earlier in Chapter 4, and the combined consequences for human life on this planet are discussed below.

- *Ozone depletion (Ozone Holes)*: the effect of ozone depletion has proved to be highly dangerous for the physiological future of the human race. It is significant that ozone destruction is the result of the same sort of polluting processes that are the cause of the other three problems; these other problems impinge di-

ISBN 1-897913-61-3. The book has a complete chapter on the Flood, in which the subject is discussed both scientifically and theologically, with conclusions drawn.

1 Zechariah 14:17

2 Amos 4:7

rectly on the weather whereas ozone depletion causes a direct and specific health hazard.

Today, medical authorities across the world report that each year there is a higher incidence of skin cancer. This is a very worrying trend which is known to be the result of increasing UV penetration which includes radiation in the damaging spectra. There have been many national campaigns to persuade people to protect themselves by keeping their skin covered by clothing and using UV barrier creams on all exposed areas. It is thought that many people are now quite well aware of the danger to themselves and their families but the increasing incidence of direct physiological problems indicates that still more needs to be done. The danger is not only the development of skin cancer but damage to eyesight and the human immune system in general may also occur.

For several decades, greater amounts of damaging UV radiation have been reaching the Earth's surface because there is less absorbing ozone in the stratosphere. This was a consequence of the chemical destruction of ozone by CFCs and similar chemicals carried up to stratospheric levels. While the countries located at the highest latitudes in both hemispheres (though more especially the southern hemisphere) are at greatest risk of the highest radiation doses, the protecting ozone concentrations elsewhere in the world have also decreased. Although these more general decreases are much less (a decrease of 5-10% from previous 'normal' values), health statistics have shown that even this small decrease has had an undesirable effect.

Through a process started by the Montreal Protocol in 1987, production of CFCs and 96 other compounds that act similarly have been stopped in most countries. Regrettably there is still some illegal production and usage. More recently, it has been noted that some new manufactured chemicals are also ozone-destroying, for example halon1202, a gas used in some fire-fighting equipment. These newer gases are not covered by the current international agreements and concern is being expressed about their increased concentration in the stratosphere; this situation is seen as a new danger for stratospheric ozone.

Finally, the role of ozone as a greenhouse gas should be mentioned again. Ozone is in fact a very powerful greenhouse gas and it has the potential to make a very small but positive contribution to the overall greenhouse effect. Restoring ozone concentrations may act to protect the human race from skin cancer, etc. but there is also a potential downside to take into consideration; the restoration of ozone concentrations will boost global warming to some degree. Current scientific opinion confirms the effect but suggests that the role of ozone in the overall global warming situation is small and may be disregarded.

- *Acid rain:* this is a factor which at first sight seems not to impinge directly on the weather. Its role in diminishing and destroying living vegetation is accepted as a major contributor to the alteration of the eco-cycle. This includes important aspects of carbon dioxide absorption and the production of oxygen. Obviously, the alteration of an area's vegetation has important implications for insect and animal life, too.

There is, however, an insidious but real link with weather and climate. In extreme cases, acid rain can cause large areas of ground to become barren or even desertified. As already discussed in Chapter 5, the character of land surfaces make an important contribution to the local weather and climate of an area. Lush vegetation exchanges heat and moisture with the atmosphere and encourages cloud and precipitation. By contrast a desert area engenders harsh conditions, arid and extreme. The change of ground cover from lush healthy vegetation to poor scrubland will certainly affect the climate of the area considerably; lesser changes in the character of the vegetation will also have their effect, sometimes more marked than expected.

Is the acid rain situation improving or deteriorating? There is no simple answer to this question. Where initiatives and agreements have been put in place, like those in North America and Europe, there have been significant reductions in the emissions that cause acid rain and a consequent reduction in rainfall acidity. Of course it is accepted that the damaging effect of the acid rain takes a considerable time to be rectified.

On the other hand the discussion of acid rain in Chapter 4 indicated how some developing countries with greatly expanding industrialisation programmes are known to be producing increasing and often uncontrolled emissions. This has led to acid rain falling upon downwind areas never before affected. So it would seem that the question of improvement or deterioration cannot easily be answered.

Most opinions tend to suggest that the overall acid rain situation is deteriorating. North American and North European reports are usually encouraging; eastern Asian figures suggest significant increase of the emissions that cause acid rain but there are some current initiatives to begin to address the problem in these areas.

- *Global dimming*: Global dimming has already been identified as an additional factor which impinges on the effects of global warming; however this is not a positive addition but a negative one. Global dimming acts against global warming; remove it and the rate of global warming would increase further. Without global dimming, the claim is that the effects of global warming would be accelerated.

  Global dimming has however a range of its own effects that alters the weather directly. These effects have been most marked within Equatorial weather regions. Here, global dimming has upset the normal climate processes that are driven by the seasonal movements of the Earth; these movements and processes are caused by the 23½ degree axial tilt of our planet as it makes its yearly orbit around the sun.

  The reduction of the sun's radiation by global dimming means that the Equatorial rain belt does not move in its normal seasonal fashion. As a result, regions waiting for their seasonal rain are afflicted by drought while other regions receive considerable rain when they expect largely dry weather. Upsetting the normal Equatorial climate has implications for climates elsewhere on the globe, since the weather processes of the Tropics have influences on climates further north and south.

So what will happen?

At present, no-one really knows what the precise outcomes of all these problems will be. There have been plenty of predictions, many based on climate modelling experiments derived from atmospheric sciences. Others derive from different branches of world science. Their conclusions often differ, in detail at least, and in the range of timescales proposed for the various events. Finally, some resort more to speculation and hope. arguing vehemently that no change will occur, or that anything that can be done now is 'too little, too late'.

Of course the passage of time will bring more certainty. More information will become available and allow predictions to be refined. This has already happened to the IPCC reports over the years. Climate models will become even more sophisticated. Computers will become faster and even more powerful.

A 2006 initiative invited the general public in the UK to run a climate modelling program on their personal computers. Those who volunteered to participate downloaded a simple climate model program which then ran in the background of their computer system and sent back its results to the climate coordinators of the experiment (the UK Met Office and several universities). Because each version of the model was slightly different, a vast ensemble of solutions was produced; these are being studied to help decide the most likely outcomes. With over 200,000 PCs involved in 2006, the effect is that of a truly vast supercomputer! To date (2010) it is reported that over 600,000 simulations have been completed and the experiment continues.

Meanwhile some voices insist that it may be detrimental, even catastrophic, to wait. They say that if the climate of a region was altered, it is certain it could not be easily restored to its former condition; it may well be impossible to restore within current life-spans. In any event, the prediction of more occasions of severe weather stands; this is the avowed view expressed by the IPCC and others.

There are also claims that the increase of severe weather events is already under way but the review of severe weather statistics presented in Chapter 5 did not provide conclusive confirmation. Although ozone depletion has been omitted from the climate-changing factors mentioned above, it may also have a very small role – but it should not be forgotten as a great physiological danger to the human race.

Nevertheless there are aspects of the warnings which are wholly realistic. A warming sea means a greater volume of water and the partial melting of the polar ice fields will continue to cause significant rises in sea levels. Pessimistic predictions suggest the sea level could eventually rise several metres but most voices argue for much less. Today, predictions of the sea level rise by 2100 are generally presented in a range around half a metre. It should be noted that even a fifth of this value is very serious, because it would affect many coastlines and cause a great loss of land to the sea.

Updating the blame issue

In Chapter 3, the conclusion was that human development had unbalanced the carbon cycle so that it is no longer self-adjusting. The conclusions of the current decade have indicated strongly (some say unquestionably) that an important and significant part of global warming has been driven by the diverse polluting actions of humanity. There are still dissenting voices but today they are in the minority.

The IPCC and the consensus of world scientists now suggest a range of 'facts' for humanity to consider. They say that increased human pollution has added to the increase of greenhouse gases. They identify the most significant pollution is linked with carbon dioxide, whose concentrations in the atmosphere have increased. The increase in carbon dioxide has been a major factor in global warming and in consequence air and sea temperatures have risen.

Although there is undoubtedly a natural warming component in the temperature rise, the human contribution is judged to be crucial, especially in the last 50 years or so. All these factors are likely to make individual weather events more extreme; we are also living more dangerously since the same factors also have the potential to produce sudden climate change by changing the sea current patterns and altering oscillation events.

It is also obvious that global dimming has further complicated the overall situation. Although global dimming acts contrary to global warming, it has its own range of serious weather-altering problems which cannot be ignored. A further complication has been the industrial pollution that causes acid rain and this continues to be a serious problem. There are damaging implications for the bio-cycle as well as the weather of the regions affected. Bio-cycle alterations are of first importance to the human race, since they impinge on the very fundamentals of life – our life.

So what have we been doing? We have all been using energy. Lots of it, for a long time. However there are now new realities in the requirements for energy which will affect us all directly. The explosion of industrial development in Asia, India and elsewhere has hugely increased global energy demand. In addition, the world population has more than doubled in the past 50 years and there is the development of considerable affluence in many regions; this has already boosted personal energy requirements greatly. It is a world ambition to eliminate poverty and improve the lot of those who live in precarious and primitive conditions; this is admirable, proper and highly desirable but it will inevitably boost energy consumption to ever higher levels.

The current major means of global energy production is the burning of fossil fuels. This directly generates the pollution that causes all the weather and climate change factors. More energy production of this type means more pollution, more carbon dioxide, more global warming, more global dimming and more acid rain unless effective action is taken.

8

Global Energy Production

It is obvious that today's 21st century world needs to produce a very large amount of energy. The total energy production is often expressed in BTUs (British Thermal Units are a unit of measurement that links energy input with temperature rise). The total world energy production value is an extremely large number of BTUs, expressed numerically in 'quadrillions' (a quadrillion is 10^{15}), with Quadrillion BTUs referred to as 'Quads'. For some illustrations, world energy is given in other units such as the electrical measurement kilowatt hours (kwh),'barrels' of oil, cu.ft of gas or tonnes of coal. In all cases the numbers generated are immense. Most calculations place the current world energy production (and consumption) around 500 Quads. This figure is suggested by various websites which present world energy statistics.

During the last few decades, the world has come to recognise two main categories of energy production. Firstly there is the energy which is generated by the burning of fossil fuels. At present, this is the way that a very large proportion of world energy is generated. Every other method of generating energy is placed in the second category. This has long been referred to generically as 'alternative energy', alternative, that is, to fossil fuels. It will be seen that the alternative energy category contains a very wide range of other technologies.

Fossil fuels – by far the major contributor to world energy

Fossil fuels are so called because they are literally made from fossils! Hundreds of millions of years ago, the fossilised remains of many forms of plant and animal life were buried and then subjected to a great deal of heat and pressure as the Earth's crust shifted and reformed all down the millennia. The result of this awesome process is an incredible store of fuel which is rich in hydrocarbons and pure carbon.

Depending upon the composition of the original raw materials and the processes which acted upon them, the fossil fuel stores could develop as solids, liquids or gases. The solids are the many carbon-rich seams of coal which have been mined for many centuries across the world; indeed the earliest records of coal mining stretch back thousands and thousands of years.

Oil is the liquid form of the fuel which is often stored under considerable pressure in rock caverns far below the ground. We are all familiar with the dramatic sight of a gushing oil well as the oil exploration experts successfully tap into a new reserve. However oil may also be stored within porous rock; this source of oil is often found quite close to the surface. In this case, the rock is scooped out using open-cast mining techniques and then crushed to obtain the oil. Finally the gaseous form of the fuel, usually described as 'natural gas', is held in huge natural reservoirs, sometimes under pressure. The gas, which is primarily methane, is extracted by drilling down into these reservoirs.

Of the three basic forms of fossil fuels, coal is the one which is most used in its raw form. Coal is burnt in furnaces and the heat energy released is used directly for heating or indirectly to provide energy for other purposes. Crude oil is refined and processed into various petrochemical fuels like petrol, diesel fuel and kerosene (paraffin) which are the fuels used in many forms of transport. The raw natural gas is processed into refined methane and it is also the origin of other fuel gases like butane and propane (generically called LPG); all these gases are used widely for energy creation in commerce, transportation and in the home. Finally, all forms of fossil fuels are also used to generate electricity.

When current world energy production is analysed, it is clear that the burning of fossil fuels is the most common method of generation; the proportion of the total is at least 80%. Unfortunately, the burning of fossil fuels produces considerable carbon dioxide and atmospheric carbon dioxide concentrations rose for much of the last century. During the last 90 years, more carbon dioxide than can be absorbed by the world's natural compensating processes has been produced.

This has been judged to be a major factor in the global warming which has occurred. Of course carbon dioxide is not the only effluent produced when fossil fuels are burnt. The smoke from each process is a complex mix of chemicals and some of these have other implications for the weather of the world – for instance, the problems of acid rain and global dimming that were presented in Chapter 4.

There has always been a desire to use alternative methods of energy generation, especially where these are thought to be 'free' and 'clean'. Some of the alternative energy generation methods have been in use for many centuries, even millennia. Since fossil fuel energy is now associated with such serious global problems, the desire to use other methods has become much stronger and alternative energy technology is now pursued with much enthusiasm and inventiveness.

Alternative energy in use today: the minor partners in global energy creation

The non-fossil fuel options have been generally referred to as 'alternative energy', although the description 'renewables' has recently gained greatly in popularity. The point here is that fossil fuels, however cheap and efficient, are finite resources which will run out some day. The most established alternative energy methods are used to produce some commercial energy. However some of the newer alternative energy methods are still experimental and not in commercial use.

It is significant that many forms of alternative energy have links with weather and climate; some are actually driven by elements of weather while most have some elements of weather dependency built in. The direct association is very obvious with wind power but it will be seen that weather links with other forms of alternative energy exist in a more indirect form. In fact, few alternative energy sources are entirely independent of weather factors. That said, nuclear power is a significant exception and is a very important alternative energy source.

Nuclear power

Generation today

Across the world, there are mixed feelings about the production of electricity from nuclear power. When commercial nuclear power plants were first commissioned in the 1950s, they were hailed as the saviours of the world. Here was clean, cheap energy produced by the very forefront of modern science. Today's picture is very variable. Some countries have embraced nuclear power enthusiastically while others have largely given it up because of the perceived dangers associated with it. Nevertheless, according to website nuclear energy statistics, around 15% of world electricity is currently produced by over 400 nuclear power stations across the world.

The majority of nuclear power stations produce electricity by a process of nuclear fission within their reactors, using a specific form of purified uranium as fuel. This reaction involves only small amounts of uranium and the process produces an incredible amount of energy, manifested as heat. This heat is then used to produce steam that powers turbines to generate electricity. It is a totally 'clean' process, with no polluting effluents.

However there are negatives. The spent uranium is highly toxic and remains very dangerous for thousands of years. Also, there are potential dangers associated with the nuclear power stations themselves – in the worst cases, serious accidents or failures can release radioactive dust into the atmosphere. When this happens, there is a direct link with weather, because the spread of the danger is determined by prevailing wind flows and other atmospheric factors.

Meteorological authorities worldwide are constantly ready to provide advice on the spread of contamination plumes in the event of accidents. In 1986, the well-known Chernobyl (Ukraine) nuclear site provided an extreme example of contamination when one of its four reactors exploded. Although thousands of deaths were predicted for the decades to come, happily there have in fact been less than 100, although it is clear that the health of the local community has suffered significantly.

There may also be dangers in the transportation of spent radioactive fuel from power stations and the long-term storage of this dangerous material is a continuing problem. Finally, although the actual energy production itself may be elegant and relatively cheap, the power stations are not – they need to be constructed and maintained to the very highest standards. So the generation of energy by nuclear fission produces its own form of pollution – not gas effluents like fossil fuels but quantities of highly dangerous material that remains so for a very long time.

Is there a better way?

In recent years, a good deal of enthusiasm has been expressed for the use of a different fuel source in nuclear reactors. This fuel, namely thorium, has been hailed by some as the solution to the dangerous waste problem that is such an integral part of uranium-fuelled generating systems.

Thorium is a naturally-occurring slightly radioactive metal which is extracted mainly from the phosphate mineral monazite. As a nuclear fuel, it is more abundant than uranium, arguably more efficient in operation and produces less waste. Although the spent thorium is radioactive, the decay of radioactivity is much more rapid than uranium, becoming safe in hundreds of years rather than thousands.

In fact thorium nuclear technology is far from new. It has been used experimentally in a number of reactors around the world for the last 50 years; in addition, some of these reactors have actually produced commercial electricity for several decades. However the benefits of thorium have been much reduced by serious technical and cost problems, some not solved as yet. The research continues. Today, there is general agreement that thorium-fuelled power stations have considerable potential for the future of nuclear power generation across the world, provided the outstanding problems can be solved.

Hydropower

Hydropower, the energy that can be captured from flowing water, has been in existence for many thousands of years. Ancient Greek literature refers to water wheels used in grinding mills, with references dating back to 4000BC. In the simplest form of hydropower, a wheel fitted with paddles is placed in the flow of a stream; the pressure of the water on the paddles turns the wheel and this energy is used to drive a range of machinery. Ancient water wheels are found all over the world and history provides reference to many more examples down the centuries. The technology is still in use today. In appropriate situations, modern forms of the water wheel are still used to power machinery directly although the main use of hydropower today is to produce electricity.

The countries that produce the greatest amount of hydroelectricity are those with substantial areas of mountainous terrain and a good deal of rainfall. The powerful streams and rivers that flow down their steep slopes can be channelled through hydroelectric turbines, where the water gives up some of its energy to drive turbine blades placed in their path. Apart from construction and maintenance costs, this is free energy; if the flowing water is not captured in this way it expends its energy uselessly, from mankind's point of view.

Globally, Canada produces the greatest amount of electricity from hydropower, exporting to the USA what it does not need for itself. The USA has its own hydroelectricity plants that provide around 10% of the nation's electricity. Many other countries with suitable topography generate significant power from hydroelectricity, including Brazil, China and India. In Europe, the mountainous country of Norway generates 99% of its electricity from hydropower. These figures were obtained from website statistics.

Hydropower is rightly regarded as a leading source of renewable energy and there seems to be very few drawbacks in its use. This is certainly true when the power is generated from constantly-flowing natural streams. Of course this technology has an obvious and direct link with weather. Where hydropower turbines are placed in natural streams, their output is dependent on the water flow through the turbines. Since the flow rate of most streams is likely to be affected by the rain that falls in the area, weather directly affects the power output of the generating plants.

In the conventional sense, hydropower is regarded as pollution free. However there is one aspect of 'pollution' worthy of mention. This applies when dams are built to capture large volumes of water in large reservoir lakes so that there will be a constant water flow for the hydropower generators. The action of flooding large valleys is one that has been repeated many times across the world and this is invariably unfortunate for the former inhabitants of these valleys; they are compelled to move elsewhere as their land is flooded to great depth. Personal hardship is imposed on these people for the 'greater good'.

Although such large scale flooding is not 'pollution' in the accepted sense, it is certainly a negative factor of this technology. In addition, the huge concrete dams that contain the water may be considered to spoil the natural beauty of an area. To counter these negative factors, it is sometimes suggested that the artificially created lakes present a valuable new source of recreation, able to be enjoyed by a great many people.

Wind power

The windmill is another ancient source of energy; history suggests windmills may have originated in ancient Persia. The word 'windmill' gives a good indication of their historical use – many were mills to grind grain harvests. Some were also used to provide power for other tasks, like drawing water from a well to irrigate crops. The link with weather is obvious and direct. Wind[1] is the

---

1 Appendix B: Wind see p. 220

generated movement of the air that encases the world, caused by differential heating in various scales. Fundamentally, wind energy is a form of converted solar energy.

In the modern world, most windmills have been renamed 'wind generators' or 'wind turbines'. The modern form of these machines uses a large rotor that is more akin to an aircraft propeller than the sails of the traditional windmill. In fact smaller wind generators may look rather like an aircraft with a large propeller – the body and tail of the 'aircraft' keeps the rotor turned towards the wind. These smaller machines have been in existence for electricity generation for many years but it is only in the last 30 years or so that the technology has been developed as a contender in the electrical energy sector.

Earlier commercial wind generators had rotors up to 10m in diameter but continued development has now expanded rotor size considerably, with some commercial land-based generators now fitted with 100m diameter rotors. The rotors in offshore installations are even larger. Research has shown that larger rotor sizes are much more efficient in terms of power output.

However the output of wind generators over any period of time will never be 100% of their capacity. As the wind speed decreases, the rotor slows and its electricity output is reduced; when the wind becomes calm or very light, electricity production falls to zero. This is why modern wind generators are carefully designed to be efficient electricity producers over a wide range of wind speeds.

So, just as there is great variability in the wind, there is great variability of output from all wind generators. This means that the generators need to be sited carefully, ideally in windy locations. This is why onshore generators will usually be found on high, exposed ground, ideally where the topography does not cause too much gustiness or turbulence in the wind. Large variations of force caused by turbulence can damage the rotor.

For all the reasons given in the last paragraph, offshore construction is desirable; the surface wind over the sea is likely to be stronger and less turbulent than that over land because there is less frictional drag on the wind over the sea. On the other hand, offshore installations are much more difficult and expensive to install and maintain, especially in deep water. Nowadays, it is common to install 'wind farms' consisting of many individual generators spaced out over an area of ground or water.

Worldwide, many countries use wind power to supplement their electricity generation. Historical pictures of the Netherlands often show a land of windmills. Today, the Netherlands is one of the nations developing wind power but their present capacity is much less that other adjacent countries; Denmark generates over 3 times more electricity by wind power than the Netherlands and Germany 15 times more.[1]

There is obviously no traditional pollution from wind generators. In the wider sense of the word 'pollution', a few negative factors can be recorded. Of these, visual impact is probably the greatest. Modern wind generators, however elegantly styled, are very large and tall objects that are likely to be intrusive to the 'view'. A large number spread across an area will be even more intrusive. This can be improved by careful siting and this claim is made invariably for all wind farms. Close-up, they may also be noisy although latest models are engineered to be as quiet as possible. There is also a degree of danger with such huge blades scything through the air! This can certainly be hazardous to flying wildlife and may also be a danger to very low flying aviators, such as paragliders and those who fly microlight aircraft.

Wave power

Ocean waves contain considerable energy and there are many places each year where great destruction is caused by powerful waves attacking coastal defences. Even relatively small bodies of water can produce waves that may be the cause of similar, though lesser, damage to containing banks or structures. For many years,

---

1 Information from website statistics.

engineers have attempted to find efficient ways of capturing this energy but the task is not so easy.

With wave power, there is a direct link with weather. Waves form on the surface of any body of water by the transfer of wind energy into the water; the stronger the wind and the longer it blows across the water, the bigger the waves become. In very large bodies of water (the seas), there is an additional type of wave called 'swell' waves. 'Swell' is the term for the remnants of wind waves generated far away. These are very long flat waves spaced far apart.

So the waves affecting coastal areas are usually a combination of locally-generated wind waves and swell waves from much further away. This is why there are always wave motions in the sea even when the wind is calm. Large sea waves can also be formed by earthquake or landslide events; the largest of these are catastrophic, as shown by the Asian tsunami of 2004. Of course, earthquake or landslide events are not part of any attempt to capture energy from the sea.

Probably the simplest and most compact form of wave generator involves the construction of a chamber which allows waves to flow freely in and out at the bottom. When the wave enters the chamber, air is driven out through an opening towards the top; when the wave withdraws, air is sucked back in through the same opening. The opening is fitted with a bladed turbine and both wave phases turn the turbine and generate electricity.

There are problems associated with siting and design; the generator needs to be placed where waves are constantly available and the design has to cope with a wide variety of wave conditions. Small waves still have energy to give. There are other methods that involve systems of floating booms, linked together and moving with the waves; the constant movement between them is translated into electrical power.

Estimates of available global wave power identify higher latitudes and western continental coasts as the best locations. Western Europe is considered to be a very good location, since it presents a very long coastline to the power of large wave-generating Atlantic storms plus the effects of swell waves from afar. The western coasts of North America also score highly, especially western Canada, while significant waves are also recorded in southern Chile, southern parts of Australia and southern New Zealand. There is still considerable research being done and the present contribution of wave power to global energy is small.

Clearly, there is no pollution involved. The energy produced is clean and free once the generators have been constructed. Such construction is expensive and obviously there are maintenance costs, also expensive. However there are few real disadvantages. Most generator installations are relatively small but those involving floating booms take up significant surface areas of water and could be a potential hazard to shipping. The energy provided by waves is generally more reliable than that of wind; even when the day is fine and calm, waves generated from elsewhere are still available to provide power.

Tidal power

Tidal power is a relative of wave power where the ebb and flow of the tide is used to drive a turbine generator and so produce electricity. Tides produce very reliable water movements with two high tides and two low tides per day - actually the 'day' cycle is almost 25 hours.[1] Power from tides can only be generated for a maximum of 10 hours per day, this being the total period of time that there are useable flows.

Tidal power stations located at major river estuaries or similar inlets involve the building of large dams to trap the tidal water and channel it through the turbines. This is a major construction undertaking and involves great alteration to the appearance and ambience of the estuary. For these reasons, the construction of such stations is rare – in the early 21st century, there was only one in operation in the whole of Europe (in France); a few others are under construction.

---

1 Appendix B: Tides see p. 214

An alternative is to locate the tidal power stations offshore. Here the technology is completely different with large rotors built in the water. The flow and ebb of the tide turns the rotors and generates electricity. Again this will only occur for a limited time each day. This is experimental technology at present although results are encouraging.

Solar power

Mankind has used the power of the sun for millennia. The sheltered sunny spot has always been sought out for pleasurable relaxation. The direct rays of the sun have been used for many purposes, heating or drying in the home, farm or workshop. The ancient Greeks and others harnessed the power of the sun through a lens to produce fire. Later, the sun's rays, reflected by mirrors, were used as an effective way of transmitting messages.

– *Solar panels:* solar panels have long been a common sight on the buildings of many hot and sunny countries. Traditionally used for heating water, the basic solar panel is a slim flat box containing a matt black heat exchanger covered by clear glass. The unit is mounted at an appropriate angle to catch the sun's rays for the maximum period of time. The sun shines through the glass, the heat is absorbed by the matt black surface of the heat exchanger and the air between the tank and the glass also becomes very hot (just like the air in a greenhouse).

  When cold water is pumped through the system, it becomes hot and is stored in an insulated tank for later use. Today, this simple type of solar panel is still installed; however, in some colder climates the system heats an antifreeze liquid and it is this hot liquid that heats the water indirectly. For many years it was assumed that solar panels were suitable only for hot and sunny climates. Nowadays, the technology has been refined and modern solar panels work well in much less sunny areas.

  A second type of solar panel produces electrical power directly. Normal daylight is sufficient to allow a photoelectric cell to produce electrical power. So today there are solar panels where the sun's energy in the form of light is captured by an array of photoelectric cells. These generate electricity which can be used directly or stored in batteries. This is the technology used commonly to power small electronic devices, like hand-held calculators. Solar panels are also very important in space technology, where they provide the power for many satellite systems.

Today, these 'photovoltaic' solar panels are becoming much more popular because the system may be linked with domestic or commercial electricity provision. This elegant system works like this: The solar panel electricity is used directly when available, augmented where necessary by the mains supply from the electricity provider; however, when the local solar power units produce a surfeit of electricity, this is automatically 'sold back' to the mains electricity provider.

Clearly, both types of solar energy panels are totally clean in the pollution sense and, provided the panels are not too intrusive, disadvantage-free. However, current installation costs are high – though these are likely to decrease with economies of scale. This is very much of a developing technology and time will no doubt bring more efficiency in terms of power output and construction of equipment.

– *Solar furnaces*: in recent years, science and technology have returned to the ancient method of concentrating the sun's rays by mirrors to produce extremely high temperatures in what has become known as a 'solar furnace'. A solar furnace positions a large array of mirrors so that each one reflects the sun's rays into a single receptor, where extremely high temperatures are generated. Temperatures over 3,000 DegC (5,430 DegF) can be generated in this way. The extreme heat is then used to generate significant amounts of electricity. This system can also used to make hydrogen fuel; here, the very high temperatures are used to split one of the hydrogen compounds.

Solar furnaces may generate electricity in two different ways. These are the photovoltaic method (which is sunshine on photoelectric cells as described in 'Solar Panels' above) and the thermo-electric method which produces steam to drive turbines. The latter method can generate considerable amounts of electricity; today, a single large system installed near Seville, Spain generates 11 megawatts of power, sufficient to meet the needs of around 6,000 homes. Large solar furnace installations are being constructed in various locations in Southern Europe, especially in

Spain, Southern France and Southern Germany. Some of these are still experimental but a few have begun to provide electricity for public use. To be optimally efficient, these installations need to be located in very sunny areas.

Like solar panels, solar furnaces generate clean, pollution-free energy. The large-scale installations are very expensive and complex to set up; also, they occupy considerable areas of ground and those areas are likely to be unusable for any other purpose.

Geothermal energy

The centre of our Earth is a huge volume of molten rock with a core temperature estimated at 5000 DegC. We live on a relatively thin surface crust that has cooled and solidified. From the surface, there is a temperature rise of 1 DegC for every 36m of depth; this is why deep mines are very warm places.

There are many places in our world where the geological structure of the area brings the molten core much nearer to the surface. These are places of great drama, where one may walk between pools of boiling, steaming water or boiling mud. Other places have the considerable drama of geysers that periodically send fountains of boiling water high into the air.

Perhaps 'Old Faithful' is the most famous geyser in the world. This geyser is located in Yellowstone Park, Wyoming, USA and it is known throughout the world for providing a very dramatic display. The name 'Old Faithful' refers to the fact the next eruption of boiling water and steam can be predicted quite accurately. Still others places in the world have hot springs where comfortable bathing may be enjoyed in freezing weather conditions. Of course, volcanoes are another demonstration, this time a very violent one, of the molten core breaking through the Earth's crust.

For many thousands of years, people have used the natural benefits of geothermal energy. The ancient Roman and Greek cities invariably included baths and central heating systems provided by geothermal energy if available. The American Indians and the New Zealand Maoris are known to have used the natural boiling waters for cooking, too. This is another example of 'free' energy, though some areas suffer from gaseous emissions that add to atmospheric pollution; however, this natural pollution would occur whether the energy was being used or not.

In the last century, domestic and industrial heating has been provided commercially by nearby geothermal energy, where available. Many suitably located homes and businesses are heated in this way. About 100 years ago, electricity began to be produced by using geothermal energy to drive turbines and this technology has been refined over the years. Today, where there is a good supply of geothermal energy, the generation of local electricity is common. Overall, however, geothermal energy makes a very small contribution to global energy production.

A more recent application of geothermal energy is the use of geothermal heat pumps (GHPs). This technology is not confined to volcanic or hot springs areas; it is a system that may be used anywhere. A GHP system makes use of the fact that earth temperatures just a few feet below the surface are far more stable than the air temperatures above. While air temperatures fluctuate widely through the seasons, months and days, the ground temperature a few feet down varies between 7 and 13 DegC. So in cold weather, the relatively warmer ground water may be used to raise the temperature in nearby buildings and in hot weather, the same water can be used to help cool them. Hundred of thousands of buildings currently make use of this type of system which is very simple and economical to run.

Biomass energy

Biomass energy (or bioenergy) was traditionally the burning of wood or other combustible material for cooking and heating; this type of energy has been in use since Man discovered how to make fire. Used in the traditional way, bioenergy is not pollution-free - but today's bioenergy is rather different. Firstly, it involves the burning of many animal and plant waste products which may otherwise go to landfill. Secondly, the burning process is carefully controlled to produce little pollution. Thirdly, it disposes of waste products that otherwise would be left to rot, when they would produce methane and other gas emissions to pollute the atmosphere.

Another initiative grows plants for bioenergy – fast growing trees, for instance. When harvested for conversion into energy (by burning), these are replaced with new trees or other plants. Thus, the carbon dioxide released during the bioenergy production process is balanced by the extra carbon dioxide absorption potential of the new plantings. The operators of such schemes claim that there is no net increase of atmospheric carbon dioxide.

There are however other ways to produce bioenergy. Waste vegetable and animal matter can be turned into fuels such as methanol, natural gas or oil by fermentation processes. This fuel (biofuel) may then be used to generate electricity. Waste oils, such as used cooking oils, can be cleaned and reused as fuel for diesel engines. In addition to providing oil-based fuels which replace the fossil fuels that would otherwise be burned, all these methods are a very efficient way to dispose of many types of waste.

Bioenergy is an energy source available all over the world for which there is an abundant supply of fuel. The world produces vast amounts of waste products and many of these are handled extremely badly at present with consequential impacts in atmospheric, soil and water pollution. Bioenergy techniques are a way to address at least part of this huge problem and in addition contribute to world energy production.

Furthermore, there is now considerable worldwide activity to produce biofuels directly from field crops; such crops are not cultivated for food but for biofuel production, for use in both petrol (gasoline) and diesel engines. The plant-based biofuels may replace completely the use of fossil fuels in both types of engine but today (in 2010) many biofuels are blended to some degree with their fossil fuel equivalent.

The production of biofuel for petrol engines uses plants which yield sugars or starch. These are crushed and fermented into ethanol, a form of alcohol. Typical examples of crops used for this process are wheat, corn and sugar beet. Biofuel for diesel engines uses a wide range of crops, the seeds or fruits of which may be crushed to obtain vegetable oils. Examples of these crops are sunflower, palm and rapeseed oils.

One of the latest forms of biofuel is algae or algal fuel. Algae grows many times faster than the most efficient biomass field crops and produces huge yields by comparison. Like all living plants, it absorbs carbon dioxide and produces oxygen by photosynthesis. Up to 60% of the oily biomass which remains after the photosynthesis process can be processed and refined into biodiesel.

Incredibly, the carbohydrate content from the remaining green waste can then be fermented into ethanol and butanol, both of which can be used in petrol (gasoline) engines. So algae fuel is very productive in terms of crop volume and each crop can provide a double output. Unfortunately, the biofuels produced by this method are currently very expensive; understandably, considerable development work continues.

Energy production stewardship: the Christian perspective

The Bible tells us that God created Man to have intelligence and free will – from which it follows that humans are not compelled to follow strategies from elsewhere, even if these strategies come from God himself. The means of making fire and all that followed from that is an excellent example of Man's intelligence. The use of wood as a combustible material was a clever strategy to solve a fundamental problem. This was Man making use of the considerable talents God had bestowed upon him.

So as the millennia passed, humans discovered other more convenient and sophisticated ways of generating heat to order, demonstrating intelligence once again. In more recent times, although the energy requirements of a rapidly-increasing world population has required increasing amounts of fossil fuel to be burnt, modern Man invented scientific methods to increase the efficiency of the processes. As evidence of this, today's fossil fuel power stations are increasingly efficient and much less polluting than those of the past. So, although it is true that polluting the Earth is against God's stewardship instructions, today's fossil fuel energy generators could claim that they are applying 'fossil fuel stewardship procedures' as far as they possibly can and furthermore that they are getting better all the time!

So what of the 'alternative energies'? Several millennia ago, wind, water and solar power was captured (to some degree) to provide energy; geothermal energy was also used where available. Today the methods of generating all these energies are more efficient and all continue to be clean in the non-polluting (atmospheric) sense. Wave and tidal power were developed much later – these too are non-polluting energy sources. Biomass energy claims to be non-polluting since the effluents are absorbed by new plantings. All these methods are completely in accordance with God's stewardship instructions - but what about nuclear power, by far the greatest contributor to the alternative energy spectrum?

It could be (and is) argued that the use of nuclear technology to produce 'clean' energy is very much in accordance with the stewardship responsibilities of mankind. Furthermore the development of nuclear power is surely an excellent demonstration of the considerable intelligence and inventiveness of mankind.

Those who are opposed to the generation of nuclear power insist that the waste products of the process (highly radioactive spent fuel) represent pollution of the worst type. Those in favour disagree strongly and argue that the nuclear pollution situation is not in the same category as the all-pervading atmospheric pollution of greenhouse gases, etc. Furthermore, they can point to alternative nuclear fuels, for instance thorium, that may reduce the spent fuel problem considerably.

Certainly, nuclear waste does not have the implications for global climate change that are associated with fossil fuel generation. However efficient, fossil fuel power stations produce carbon dioxide and other greenhouse gases. This fact is a powerful and compelling argument for the retention of nuclear power generation.

In the 21st century, many countries have a rekindled enthusiasm for nuclear power and some have well advanced plans to construct more nuclear power stations in the next decades. No doubt these will be constructed using the latest techniques to minimise the disadvantages and dangers. So it would seem that God's stewardship requirements are being observed by those in favour of nuclear power generation – mankind using human talent, intelligence and free will once again.

9

Global Energy Consumption

The Creation... of an energy consumer

In the Creation story of Genesis Chapter 1, the Bible is unequivocal about the food that is to be eaten by a newly-created mankind. Having given his 'finest creation' dominion over the world including all its plants and animals, God then specified that humans should eat only fruit and certain vegetables.[1] Likewise, all animals were to be vegetarians also.[2] As you may imagine, these texts have long been popular with the followers of vegetarianism!

However, this vegetarian diet was specified for mankind in their 'Garden of Eden days' before disobedience brought about 'the Fall' and a rapid change in their life conditions.[3] In other words, the fruit and vegetable diet was for men and women in

---

1 Genesis 1:29
2 Genesis 1:30
3 Genesis 3: all verses

their 'pure' state. After the Flood, the food specifications were radically amended.[1] Just before God announced the terms of the Rainbow Covenant, he gave Noah and his family permission to eat animal life of all types: 'Everything that lives and moves will be food for you.'[2]

So from that day, mankind was authorised to obtain personal energy (food) from unrestricted sources, although it is of paramount importance to note that their stewardship responsibilities for the Earth remained unchanged. Furthermore, there are references to cooking meat and fish throughout the Bible; the earliest are found in the book of Exodus when God was issuing instructions about actions and behaviour.[3]

The rest of the Old Testament and the complete text of the New Testament do not alter God's dietary permissions for his people. There are many references to the eating of meat and fish. Jesus himself is recorded eating meat and fish. The Last Supper with Jesus as the host was the Jewish Passover meal at which a sacrificed lamb is eaten. [4] On another occasion, Jesus attended a rich banquet at the house of Levi, a tax collector;[5] no doubt the meal included meat and fish. Finally, the resurrected Jesus ate some fish after appearing to his disciples.[6]

Man discovers how to make fire

The human body (prehistoric or modern) needs to maintain internal energy levels to continue to live. Basically, the fuel for life is food and water. In addition, the sun provides essential heat but this is variable by day and non-existent at night, although the heat of the previous day helps to keep night-time temperatures from becoming too cold. So humans, lacking the natural protective coat of many animals, have to maintain their essential body environ-

---

1 Genesis 6-7: All verses
2 Genesis 9:3
3 Exodus 12:8 (roasting), Exodus 23:19 (cooking)
4 Mark 14:12-17
5 Luke 5:29
6 Luke 24:42-43

ment by wearing clothing and seeking shelter from the elements when conditions become inclement.

In the far distant past, prehistoric Man would be likely to choose to live in an area where he could obtain fresh water; this would be supplied directly from rivers or from the collection of rain. These water sources would also ensure that a range of food was available. Trees or bushes would be a source of fruit or berries. Other plants may offer nutritious root crops and their leaves may also be edible. Fish could be taken from the rivers and a wide range of insects, birds and animals could be captured and eaten as a source of protein.

The traditional picture is of a carnivore who hunted and ate red meat, birds and fish, supplementing these with fruit and vegetable plants. Since fire was not yet discovered, all these foods would need to be eaten raw. In due course, prehistoric Man discovered how to make fire and a new chapter of human life was opened. Now, humans could cook their food to make it more digestible and palatable; they could also warm themselves when the sun was absent and so conserve their personal energy. This was a very significant development for another reason, too. For the first time, Man became a consumer of fuel energy and a polluter of the atmosphere.

Today's energy consumption flows directly from these humble beginnings. For very many generations, people could only create heating energy by burning wood or similar products. Finally, around 10,000 years ago, coal was discovered and found to be a very convenient and efficient fuel source. The use of coal as a fuel source started the fossil fuel millennia. Much later, the discovery of oil and gas followed. As the world population increased, the use of fossil fuels proliferated and the cumulative effects of these are now evident for all to see.

Analyses of energy consumption today

Many reviews of worldwide energy have been published over the years. Production is of course driven by consumption with only temporary imbalances between them at times. Production data is updated constantly and is distributed widely in scientific circles;

nowadays, the figures are published on the Internet and (when sufficiently dramatic!) appear in various ways in the media.

Scientific reviews of energy usage are full of statistics that aim to rationalise and break down the huge global consumption figures into meaningful divisions. All the figures that appear in this chapter were obtained from a range of websites which review and present such statistics. This information can be used to address two important questions which are of fundamental importance in the arena of global warming and climate change. These refer to the 'what' and the 'who'. The 'what' refers specifically to activities, identifying the areas of human activity responsible for using this energy across the world. The 'who' identifies the location of the energy users regionally and nationally.

'What' is using all this energy?

Analyses of energy and world activities are usually presented within named 'sectors' of energy consumption. These suggest that 'industry' is the biggest user, responsible for consuming around half of the world's energy production. This is hardly surprising since virtually every manufacturing or industrial process is a heavy user of energy. Second in line comes 'transport', consuming an estimated 15% slice of world energy. The energy consumption of the 'residential' sector is not far behind 'transport'. The remaining balance of world energy production is used up by a range of proportionately smaller sectors such as 'commerce', 'agriculture' and the energy usage of 'energy production' itself.

A large number of high energy consumers use primary energy sources – commonly one or more of the fossil fuels. Oil and natural gas are efficient and very easy to use for many purposes while coal, although very efficient as a fuel, is more labour-intensive to handle and thus generally less popular. However any discussion of energy is unlikely to progress far without the mention of electricity. Every one of the sectors mentioned in the last paragraph uses electricity to some extent and there is no doubt that a large amount of worldwide energy is delivered by that source.

Obviously, electricity is a highly convenient and multi-purpose source of energy that is able to drive a huge range of machinery –

everything from the simplest personal electrical device to the largest and most complex commercial and industrial plants. Furthermore, virtually every technological advance involves the use of electricity as a power source.

This shows why the demand for electricity worldwide continues to soar. Particularly steep increases in electrical energy consumption are noted in developing countries where there is considerable and very rapid growth in industrial production. As a consequence, this sort of growth generates increasing affluence which in turn leads to the greater use of residential electricity to power newly-acquired domestic appliances, heating and air conditioning systems, etc.

It is important to remember that electricity is not a primary source of energy. Electrical power is generated from a range of primary energy sources, especially those which involve the burning of fossil fuels. At present, it is suggested that two-thirds of world electricity production is generated by burning fossil fuels of one type or another. So although the use of electricity as a power source appears to be non-polluting, it may well have been generated by fossil fuel burning at a power station and so it becomes indirectly polluting.

On the other hand, it is possible to generate electricity from renewable (pollution-free) sources; hydroelectricity is a significant contributor and is important in some parts of the world where sufficient amounts of suitable flowing water are available. It is estimated that over 15% of electrical energy worldwide is generated by hydropower. Unfortunately, many countries do not have the topography to support hydropower at any significant level.

Electricity may also be generated from other forms of natural power (for example sun, wind, wave) but all together these contribute proportionately little, perhaps around 3% of the world total. However there is considerable activity within these technologies and their contribution is expected to increase. Nuclear power provides the remainder of world electricity generation; the estimate is around 15%.

Like the industrial sector, the worldwide residential sector may also consume a range of primary energy sources, although electricity is usually involved, too. Houses may be heated by the burning of oil or natural gas but the means of controlling and delivering this heat is almost invariably by electricity. However the domestic energy used in many developing countries still involves simpler methods of cooking and heating, based on the burning of wood or coal in open fires.

Transportation is rather different from all the other major energy sectors. Here, almost all of the energy consumed derives from oil, that is, petrochemicals in various forms. Even when particular types of transport are propelled by 'clean' energy sources – like electric trains, for instance – the energy source is often generated from fossil fuels and thus the 'clean' vehicle becomes an indirect emitter of pollutants.

When the subject of transportation is discussed, it is usually not long before the private motor car is identified as a major energy user. In such discussions, energy-conscious citizens who take pride in their ownership of small energy-efficient cars decry the existence and use of larger, powerful cars. 'Gas-guzzlers' are identified as the villains of the piece! Obviously the motor car uses energy and is a direct source of pollution; the bigger and more powerful the car, the more polluting it is likely to be. However it is worth noting that today's motor cars are considerably more energy-efficient than those of past years and they are also considerably less polluting.

Within the spectrum of worldwide transport, it is accepted that the cumulation of private motor cars do use very large amounts of energy and emit considerable greenhouse gas pollution. However, the many other forms of transport also contribute greatly. People everywhere are travelling farther and faster around our planet. Air transport continues to grow with more and often larger aircraft taking to the sky, consuming vast quantities of fossil fuel. Public railway trains have acquired more powerful engines to propel them at higher and higher speeds. Gigantic trucks transport freight by road, covering long distances to meet the ever-increasing requirements of a demanding and growing population. Immense ships sail the seas with hugely powerful engines that consume great quantities of fossil fuel.

Within each of the sectors identified in the analysis above, energy efficiency will always be an important factor. A high level of energy efficiency in a sector will reduce its consumption considerably while poor energy efficiency will have the opposite effect. All aspects of energy efficiency will be dealt with in the next chapter.

'Who' is using all the energy?

Every country in the world consumes its own proportion of global energy and a number of organisations publish annual statistics and estimates, all available on the Internet. Of course there are great variations in national energy consumption across the world. For each country, a further breakdown of the figures makes it possible to see how the consumption total is made up from their various energy sectors.

A complex range of factors influences the energy consumption of each country. For instance the physical size and population of the country will have major influence on the energy consumption figures as will the extent of commercial and industrial activity in that country. There are also external factors that contribute. For instance, the regional and local climate will control how much energy may be used for heating and cooling; even that is complex, because the affluence of the country and its population must feature in that equation too.

Nevertheless it is easy to appreciate that large, rich and technologically advanced countries will be large energy users, while small, relatively less developed countries will be low energy users in comparison. Indeed, it is estimated that a considerable number of people in the world have as yet no access to electricity; the estimate is in the order of 1.6 billion people. Incredibly, this represents over one-fifth of the world's population. Obviously, the extension of electricity to these peoples (surely an inevitability) will add a great number of high energy consumers to the world totals.

Until just a few years ago, every calculation of energy consumption placed North America at the top of the energy consumption list by a considerable margin. Such high energy consumption was unsurprising; the USA is a large and highly-developed country with extensive natural energy resources and a considerable population. Canada is a very large country, also rich in natural resources but it is much more sparsely populated than the USA, with most of its population concentrated in the south of the country towards the USA border. In these southern areas of Canada, population density is broadly similar to that of the USA.

During much of the last decade, the second largest energy user was always identified as China. The 21st century has seen a very large increase in the energy usage of this vast country and this increase continues, due to rapid industrial expansion. In the early years of the 21st century, China's gross consumption of energy was only half that of the USA. Today, China has overtaken North America as the largest user of world energy. However on a *per capita basis,* China's usage is considerably below North America and most other developed countries.

In the past, when the former Soviet Union was listed as a single country, it was also a large energy user. Today, Eastern Europe and the former Soviet Union countries consume less than half the energy that China does. As expected, the industrialised countries of Western Europe also use a considerable amount of energy; individually, Germany has the highest proportion of Western Europe energy consumption (about one-fifth) with the UK around two-thirds of that. In comparison, the other regions of the world use much less energy although certain very affluent countries (e.g. the oil-rich Gulf States) tend to have very high *per capita* consumption values.

In fact the *per capita* consumption values show a link between the location of exploitable energy resources and the energy usage at that place. When a country is rich in energy production its population tends to be affluent and the *per capita* energy consumption value is correspondingly high. The highest *per capita* values are found in the rich Gulf States where the largest reserves of oil are located and where there is considerable production of that energy source. The citizens of these countries have a very high standard of living which in today's world means that a great deal of energy is consumed; in addition, the climate of the region means that considerable energy is expended on air conditioning and water delivery.

Interestingly, the affluent countries of the far north (e.g. Canada, Iceland, Norway) have also revealed high *per capita* energy consumption; like the Gulf States, the citizens of these countries have high standards of living and the climate also requires the expenditure of considerable energy – but this time usually for heating, not cooling.

Summary: world energy production affects global warming

The direct link between world energy production/ consumption and global warming is now accepted. This is because at least 80% of global energy is produced by the burning of fossil fuels, the significant effluent of which is carbon dioxide, an important greenhouse gas. The build-up of carbon dioxide concentrations in the atmosphere have been shown to link positively with the increase of global near-surface temperatures.

However, it is striking that the 20% balance of global energy production has no implication for global warming, although bioenergy depends upon its effluents being neutralised by schemes of managed re-absorption. On the other hand, almost all alternative energies can demonstrate some disadvantages which could be labelled as 'pollution' in the wider sense but this is likely to be 'sight pollution' or 'noise pollution' etc. Unfortunately in this cost-obsessed world, all the energy produced by alternative systems is at present more costly than traditional fossil fuel production. This is due largely to setup costs and economies of scale; small-scale production invariably costs more.

As an aside, it is worth noting that the true cost of fossil fuels is actually greater than the direct cost of inputs. The difference is what economists call an 'externality': in short, the one who burns fossil fuels keeps all the energy benefit, but customarily disregards any pollution consequences, which will be borne by everyone. Various economic remedies have been proposed to modify the incentives in favour of alternatives, mostly by attaching a notional cost to effluents; for instance 'cap and trade' of carbon permits was discussed at Copenhagen 2009. The most poignant externality is that the largest energy consumers often live in temperate regions better able to face any consequences of global warming while the energy-poor often also live in regions that are semi-arid at the best of times, for whom even a small change may spell disaster…

'The foundations of the world'

The book of 1st Samuel is one place in the Bible where God's role as creator is poetically expressed. Hannah had been blessed by the birth of her son Samuel and now she offers a poetic prayer to God which included the words: 'For the foundations of the world are the Lord's, upon them he has set the world'.[1] The prayer goes on to emphasise that human strength does not come from oneself but from the Lord. There is also a similar acknowledgement of God's creative works in Psalm 89. After stating that God created the world, the psalmist goes to speak of God's attitude towards his creation. 'Righteousness and justice are the foundations of your throne, love and faithfulness go before you'.[2] These themes are repeated in a number of other places in the Bible.

It is clear that human beings have been made very finely; their intelligence, capacity for creativity and love is so obvious. The Bible expresses this succinctly when it tells us that Man was created by God as his 'finest creation' and 'in his own image'. [3] God's attitudes towards the world are righteous, loving and caring and he wants his 'finest creation' to follow his lead. The encouragement to do so is enshrined in God's stewardship instructions to mankind and these have become more and more relevant in today's 21st century situation.

As the world becomes increasingly interdependent, it becomes ever clearer that populations cannot afford to live

---

1 1 Samuel 2:8
2 Psalm 89:14
3 Genesis 1:27

selfishly, with no regard for their fellow men in the global sense. Inevitably, this consideration must direct the actions of each individual, because each action within any population impinges upon the whole world situation. This is certainly true of weather and climate change. The world has just one 'foundation' and the survival of the world and its inhabitants depend upon coordinated and unselfish action across its lands. Constant actions of love and nurture, in fact.

10

Back To 'Christian World' Basics

Dominion and stewardship

It has to be restated. Dominion and stewardship are undoubtedly two of the 'basics' of the Christian World. They are the words that summarise the authority and responsibility that mankind has been given for all aspects of the fabric of the world, living and inanimate. In the Bible, the very first reference to authority over the Earth is in the Creation story set out in Genesis Chapter 1; God said to Adam and Eve 'Be fruitful and increase in number; fill the Earth and subdue it'.[1] Stewardship is emphasised not long after:[2] 'The Lord God took the man and put him in the Garden of Eden to work it and take care of it.' Here, the Garden of Eden is representative of the whole Earth.

---

1 Genesis 1:28
2 Genesis 2:15

The dominion and stewardship of mankind also applies to the living creatures of the world, first stated in Genesis Chapter 1 'Let us make man in our image, in our likeness, and let them rule over the fish of the sea and the birds of the air, over the livestock, all over the earth and over all the creatures that move along the ground.'[1] Thereafter, these responsibilities are reiterated in many ways throughout the Old and New Testament texts.

It all makes so much sense. Indeed the human inventions of logic and reason would agree. The Earth is the place where we all live; it is the place that sustains us, giving us the air without which we could not breathe, providing us with the food and water that we need to fuel our bodies. It is filled with the awesome organisation of nature to keep its fabric in good health, to maintain its balance. Yet we can destroy its balance by casual and uncaring acts, rooted in selfishness; when we do so, we are moving towards destroying ourselves.

So it makes sense to treat our world with care and consideration so that it will continue to support us in perpetuity. It makes equal sense to look after all its living plants and animals, for they coexist with us as part of this wonderful world. It makes sense for us to look after each other, extending this not only to family and friends but to all the citizens of the world, for it is clear that we are interdependent. We are all in it together.

The power of love: The human dimension

The word 'love' is extremely important in the Bible. The Old Testament, which comprises 80% of the Bible, has around 60% of the biblical occurrences of the word 'love'. The New Testament, physically much smaller, has 40% of the total occurrences. So the New Testament has proportionately many more occurrences of 'love' within its texts, almost three times more, in fact.

Furthermore, there is a significant change in the use of the word 'love' along the timeline of the Bible. Many of the Old

1 Genesis 1:26

Testament occurrences refer to human love between marriage partners or between parents and children; however there is also a strong emphasis on the requirement of God's people to love him. A typical example of God's love is found in Exodus, as God speaks to Moses: (*God speaks*)....(I am)... 'showing love to a thousand generations of those who love me and keep my commandments'. [1] Significantly, it should be noted that this is an important statement of reciprocal love. Many other books of the Old Testament have declarations of God's fulsome love for his people.

The New Testament introduces the inspiring concept of unconditional love, a category of love shown so clearly in the person of Jesus Christ. Jesus encapsulated all his teaching about love in his New Commandment: 'Love one another, as I have loved you';[2] this commandment was very special because it encompassed and expanded earlier commandments such as 'Love your neighbour' or 'love your enemies'.

So the moral position of those who live in the Christian World is clear. They should act with responsibility towards the Earth and its creatures; they should also act with unconditional love towards all other people on Earth. Obviously, these fundamental concepts should be applied to world problems, such as those involved with global warming and the other climatic and physiological matters identified in this book.

Focussing on weather and climate issues, today's realities are these. The warming world now identified as global warming has been known about much of the last century. Acid rain has been recognised since the time of the Industrial Revolution but has only been addressed scientifically during the last 50 years or so. Global dimming is a more recent discovery; however it has been studied, understood as a dangerous effect with detrimental consequences requiring corrective action. The excessive depletion of protecting ozone concentrations in the high atmosphere has been identified,

---

1 Exodus 20:6
2 John 13:34

understood as a totally man-made problem which needs to be acted upon with urgency. Finally, warnings of the very real possibility of weather and climate change have been in existence for three decades and have been strengthened considerably in more recent times.

The laws and morals of the Christian World would surely demand action on all these problems. Has this happened? Happily, the answer is 'yes' and, importantly, these actions have involved many countries within and without the Christian World, too.

The mitigating actions of the World

The last thirty years have seen serious international attempts to identify and address the problems associated with global warming and potential weather and climate change. Many learned bodies have been set up and are still working together to identify the best ways forward; in addition there have been global political initiatives. There are also a significant number of scientific establishments working on the problems. The following chronological list shows some of the most important initiatives.

International agreements

1979 The World Meteorological Organisation (WMO), a specialised agency of the United Nations (UN), hosted a World Climate Conference in Geneva, Switzerland to discuss all the problems of climate change and explore possible ways forward. This conference was attended by scientists from all over the world.

1987 In the 1980s, there was considerable activity across the world when the physiological danger of ozone destroying chemicals was recognised. After various meetings, studies and agreements, the Montreal Protocol started the process to phase out the production of CFCs and many similar ozone-destroying chemicals. The protocol came into force in 1989 and there have been a number of revisions since then. The Montreal Protocol is hailed as one of the most successful international agreements which has been negotiated. It has 196 signatories, spread across the world.

1988 The Intergovernmental Panel on Climate Change (IPCC) was established by the WMO and the United Nations Environment Programme (UNEP) was set up to give independent advice to members of the UN on all aspects of climate change. The research into global warming and climate change is not carried out directly by the IPCC but by scientific groups across the world. The results of research are coordinated by the IPCC and reports are issued periodically. Also, the problem of acid rain was addressed in this year. 24 nations in Europe and North America signed up to a UN-sponsored agreement to reduce acid rain producing emissions, mainly identified as nitrous oxide and sulphur dioxide.

1992 The Earth Summit held in Rio de Janeiro produced the United Nations Framework on Climate Change, an international agreement designed to tackle the problem specifically by reducing greenhouse gas emissions. 172 governments participated in this summit and 165 states signed up to the Convention. This was the conception of the well-known 'Kyoto Protocol'.

1997 The Kyoto Protocol set binding and mandatory targets for each country to reduce greenhouse gas pollution. The overall aim was to reduce emissions to 5% below 1990 levels, this to be achieved by 2010. This actually represented a cut of almost 30% from the predicted 2010 levels. In 1997, 141 states ratified the agreement, representing 85% of the world's population. The USA was a notable exception; this was particularly significant because the USA then produced at least one-fifth of global greenhouse gases.

2005 The Kyoto Protocol came into force formally on February 16th, 2005.

2006 The EU began coordinating region-wide controls by carbon emission trading. The UK Government proposed legislation to drastically cut carbon emissions by a series of 5 year targets. A review of the worldwide economic implications of climate change was also commissioned by the UK government; this resulted in the issue of a comprehensive report by a leading UK economist. The report suggested that vigorous action was required to reduce carbon emissions and that the benefits of early action would considerably outweigh the cost of doing this. The report also warned that delayed action would boost costs to astronomical levels. Reactions to this report have been a mixture of praise and criticism.

2007 The IPCC confirmed their warnings and suggested refinements to their estimates. If no action is taken against global warming, the global average temperature could rise 3 DegC at least by the end of the 21st century. The sea level rise was forecast to be around 50cm by the end of the 21st century. This report was endorsed by over 100 countries. The concepts of 'carbon footprint' and 'carbon neutral' activities were introduced in many countries. People were encouraged to be aware of their personal carbon emissions (footprint) and to take action to reduce them. Many commercial companies stated that they intend to become 'carbon neutral' as soon as possible.

2008 Following a change of government in the USA, the new administration accepted the need to cut carbon and other emissions and promised a comprehensive set of actions to do so. This included the use of all energy efficiency methods and alternative/renewable energy production methods. A USA Climate Bill was constructed, proposing significant percentage cuts in carbon emissions from 2005 levels.

2009 By the end of the year, the USA Climate Bill was still under discussion in the Senate and has not been passed into law. Meanwhile the 15th United Nations Climate Conference met in Copenhagen in December, with 192 countries taking part. It was attended by the Heads of State of all major countries. The aim was to achieve an ambitious global climate agreement for the period after 2012, when the Kyoto agreement expires, with rich nations contributing financially to the development of carbon reduction strategies in the poorer nations, for the benefit of all. There was much consensus expressed across the nations and it was felt that progress had been made. However no legally binding treaty was agreed.

2010 187 countries have now signed the Kyoto Protocol. However it should also be noted that the Kyoto targets generally do not apply to 'developing' countries, including China and India, where there is very large industrial expansion taking place.

'Renewable' energy solutions in use

Today's renewable energies are another part of the mitigating actions in progress. With just one exception, all these energy generating sources are non-polluting in the fossil fuel sense; that is, they do not emit greenhouse gases. The sole exception is bioenergy but it is claimed that the carbon dioxide emissions of managed bioenergy are neutralised by the absorptions of new plant growth. In other words, bioenergy is a 'carbon neutral' process.

– *Nuclear power*: popular initially, nuclear power has tended to fall out of favour in some countries largely because of its very dangerous residue but also because of the expense of power station construction and maintenance. In addition, there has always been a fear of accidents and the threat of global terrorism is a more recent addition to the list of objections. Some countries have actively decommissioned nuclear power stations and thus reduced the totality of global energy production.

  Supporters of nuclear power suggest that if there is a serious intent to decrease the use of fossil fuels for energy production then the increase of nuclear power generation presents an almost irrefutable case. The technology is tried and tested. It is in operational use in many countries. It is claimed that new construction methods make it easier and cheaper to build new nuclear power plants; also that the use of different nuclear fuels (e.g. thorium), could greatly diminish the nuclear waste problem. In the 21st century

tury, a growing enthusiasm for nuclear power is observed. In the circumstances, this seems justified.

- *Hydropower:* a major expansion of hydropower would certainly contribute to the solution of the fossil fuel problem. There must be areas with significant natural water flow whose energy could be captured for hydropower, perhaps by smaller 'local' units. Any person or group with access to flowing water could install a small hydropower generator and thus become more energy-efficient. These installations should be relatively cheap when standard designs are developed.
- *Wind power:* those who argue the case for wind power claim that continued research will result in more efficient wind generators, and a 'minimum pollution' policy should site wind farms well away from the view of most people. There is significant enthusiasm for offshore constructions. Like hydropower, perhaps there could also be a focus on smaller units. These would be cheap to install and should not be 'sight polluting'. Those who live in windy locations could certainly benefit from personal wind generators; indeed, some people have had this type of equipment for many years. Certainly, community or personal wind generators are becoming much more popular and seem to offer a simple and potentially viable way to contribute to energy efficiency. Smaller wind generator units are now marketed in some countries for purchase by the general public, either as individuals or communities. However it is important to stress that a windy exposure is essential; it is thought that most urban areas are unlikely to be suitable.
- *Wave and tidal power:* it is obvious that the vast power contained in the movements of sea water is in the main expended uselessly on shorelines. Indeed, this natural power causes erosion and, in its more ferocious phases, great damage. Enthusiasts for the capture of wave and tidal power claim that the technology could be used much more widely and there are positive development plans in many areas of the world where tidal flows are significant. No doubt further research into large generating units will continue; however it is also suggested that many smaller installations would allow people who live near coasts to benefit from their own personal or community units.

- *Solar power:* this energy is free, renewable and non-polluting. It is easy to capture and has the potential to make a significant contribution towards the reduction of the energy drawn from fossil fuels. It makes sense that people should be encouraged and assisted to install as much solar power generation as possible. Traditional solar panels are still available; these are simple to install and provide hot water heated by the sun. However many of today's solar panels are the 'high-tech' photovoltaic type which generate electricity from light; this electricity supplements your domestic supply, increases your personal efficiency and reduces your electricity bills.

  Solar panels of all shapes and sizes are used for many scientific purposes, including the powering of spacecraft and experimental vehicles here on Earth. In addition, the solar furnace technology described earlier can be applied to small systems by employing single parabolic mirrors. This is yet another way of capturing 'free' energy.
- *Geothermal systems:* where geothermal sources are nearby, it makes sense to use them for heating or the generation of electricity, both of which may be quite easily achieved, especially if the geothermal source is near the surface. Geothermal systems usually provide considerable amounts of heat energy and can often provide energy for quite large communities.The geothermal heat pump (GHP) system is a separate technology that is not location-dependent, being based upon the more conservative temperature variation just below the ground. A simple pumping system can then provide heating or cooling as appropriate to nearby buildings. This should be designed into new buildings everywhere.
- *Bioenergy*: bioenergy not only generates usable energy for the world but helps to address the considerable problem of human waste disposal – a very serious problem. Used extensively, this strategy could be significantly beneficial to the global situation. At present, its use is quite limited.

  In fact, bioenergy projects are at the 'high-tech' end of the waste disposal industry. The benefits of recycling are well known everywhere and many countries and authorities operate schemes to reuse a whole range of materials. However the operation of recycling is extremely variable across the world and much more needs to be done to optimise the considerable benefits and efficiency gains that are available from such operations.

Other initiatives: capture and storage of greenhouse gases

To address the increasing carbon dioxide concentration problem, all the initiatives already discussed in this book have sought a solution which involves the reduction of emission levels. This has been applied to all sources that produce carbon emissions, encompassing industry, commerce, agriculture, domestic and transportation activities and processes. This emission reduction approach may be the first and most obvious option but the problem can be approached from the 'other end' too. Emitted carbon dioxide may be captured, processed and stored out of harm's way. Many propositions have been examined and experiments carried out; many more are planned.

Generally, below-ground storage has been the most popular option with deep caves and disused coal mines considered. The voids of spent oil or gas wells are another possibility and there have been experiments to examine storage of the material beneath the sea bed. Some results have been encouraging but there is always concern about leakage and what effect this may have. One way which could solve the leakage problem is to use chemical processes to combine carbon dioxide with other elements and so produce harmless solids. This research continues but its potential contribution to the problem is as yet unknown.

There is similar research on the capture of the other major natural greenhouse gases but the activity is limited. Farm animals, especially those whose digestion systems are of the ruminant type, produce considerable amounts of methane which adds to greenhouse gas pollution. It is possible to use systems to capture at least some of this methane and this has been done at some locations. Some power has been generated in this way and at least one organisation refers to this quite accurately as 'cow-power'!

Alternative fuel strategies

- *Hydrogen fuel and fuel cells:* this is another much-discussed topic in the early 21st century. Fuel cell technology already exists and is in use experimentally. In this process, hydrogen gas is combined (for instance) with oxygen in a specially designed fuel cell. The resulting reaction produces power in the form of electricity and the emissions of the process are heat and water. This is actually the reverse of the 'electrolysis' process, where water is separated into hydrogen and oxygen by the passage of an electric current through it. Electrolysis is vintage technology discovered over 200 years ago. Fuel cells may also use other gas combinations; hydrogen (or a hydrogen-rich substance) may be chemically combined with various other elements.

Hydrogen is an element that is very common in nature. Despite this, it does not exist on its own but always appears in combination with other elements. For example, hydrogen combines with oxygen to form water. So before it can be used in a fuel cell, hydrogen has to be 'manufactured' and this tends to be an expensive process. Unfortunately, the least expensive ways of producing hydrogen use fossil fuels with their greenhouse gas emission problems. Hydrogen is also the lightest element in nature. The lightness of hydrogen makes storage a problem and it is usually stored as a pressurised gas or as a deeply frozen (cryogenic) liquid; the former requires expensive pressure tanks and the latter very specialised equipment and vessels.

Fuel cells have actually been in existence for over 150 years. In some senses, they are rather like batteries but they have a much greater capacity because of the fuel stored within them. In earlier years, hydrogen power tended to be focussed on transportation. During the latter part of the 19th century, hydrogen-fuelled internal combustion engines were developed, but more recent research has concentrated upon fuel cells generating electricity and driving electric motors to power vehicles.

Such vehicles can be totally non-polluting in their operation, if the hydrogen that powers their fuel cells has been produced by a means other than fossil fuel burning. Such means do exist; it will be recalled that solar furnace energy may be used to produce hydrogen, for instance. However if the hydrogen has been produced by a method involving fossil fuels, then the vehicle becomes an indirect polluter.

On today's roads, there are now experimental hybrid-powered vehicles where fuel cell electricity is one source of power in the vehicle; the other source is usually a conventional petrol or diesel engine. The fuel cell is supplied with hydrogen stored in liquid form under high pressure; an electric motor is powered by the fuel cell and drives the road wheels of the vehicle. The petrol or diesel engine powers the vehicle in the conventional way. The two power systems may be linked together in a common transmission arrangement so that each may be used optimally. These vehicles are claimed to be much more economical than conventional vehicles and the pollution from them is reduced significantly.

Today, hydrogen power fuels space rockets and fuel cell technology has been extended to power many other types of vehicles, including at least one submarine. Several major motor manufacturers have new production facilities devoted to fuel cell powered vehicles. Although at present the production and storage difficulties remain, the enthusiasts for this technology promise that both problems will be solved in the future and that hydrogen will replace polluting fossil fuels as the major energy source.

It is suggested that hydrogen will not only power our transportation but will provide the electrical power that we require for all our normal domestic and industrial uses. The timescale is difficult to assess but a popular suggestion from enthusiasts is 'by 2050'.

– *More fuel alternatives for petrol or diesel engines*: this approach intends to cut greenhouse gas pollution by replacing petrol and diesel fuel with 'cleaner' and (ideally) more efficient alternatives. The most common solution is the use of liquid petroleum gas (LPG). LPG is the generic name for commercially produced propane and butane and this fuel has been in existence for many years. It is a by-product of methane. Traditionally, LPG has been used to provide gas for heating and cooking in those areas where mains gas supplies are not available. It is stored under pressure in cylinders and tanks that cover a wide range of volumes.

Vehicles with petrol or diesel engines can be modified to use LPG. The modification process includes the fitting of a LPG pressure tank, usually in addition to the existing fuel tank. There are of course cost implications. Once fitted, the vehicles are claimed to be much 'greener', since they produce less carbon dioxide emissions than their petrol/diesel equivalents. In the case of petrol, a 20% reduction is claimed. Other pollutants are reduced even more; Nitrate emissions are reduced greatly – up to 50% when compared to petrol engines.

Compared to conventional diesel engines, this reduction is considerable, with twenty LPG commercial vehicles producing the same nitrate emissions as one conventional diesel engine. In addition, particle emissions (a big problem for diesel engines) are cut by half. Today, some vehicles have been designed to be 'bi-fuel'; here the vehicle uses LPG until the tank is empty and then switches to petrol or diesel.

There is also enthusiasm for reprocessing used and discarded cooking oils into a fuel. While this can be carried out as a commercial process, it is now possible to obtain small, 'personal' reprocessing systems that can be used by individuals to produce small quantities of fuel. The process cleans, refines and blends the cooking oils with ethanol to produce a fuel that is usable in a diesel engine.

Other fuel alternatives are linked to bioenergy. Today, large quantities of ethanol are produced from plant fermentations, mainly corn. Ethanol is described as a 'low carbon' fuel which can replace fossil fuels in internal combustion engines. It is reported to emit 30% less greenhouse gases when compared to petrol. The blending of ethanol with petrol is a possible way forward; in some countries, this is already done to some degree.

Nuclear fusion

No discussion of alternative energy is complete without the inclusion of nuclear fusion – in a sense the 'Holy Grail' of energy. The raw materials needed for nuclear fusion energy production are plentiful and the generation process is pollution free in the fossil fuel sense. In addition, the residue of the process is harmless in comparison with that of nuclear fission; any radioactivity in the spent fuel material decays rapidly.

Nuclear fusion, as the name implies, is an energy created by the fusing together of certain atoms. This differs from nuclear fission (as used in existing nuclear power stations) where the atoms are split apart. Nuclear fusion is a wholly natural process that takes place continuously in the sun and all the stars in our universe. Within our immensely hot sun (up to 15 million DegC), hydrogen is converted to helium and the result is the energy emission that lights our world and sustains our lives.

The challenge has been to produce excess power from a nuclear fusion process and many attempts have been made to do this in the last decades. The problem is that incredibly high temperatures are required to make the atoms fuse. It could be said that one successful 'commercial' application of the technology has been the hydrogen bomb. This is indeed nuclear fusion but it is totally uncontrolled – suitable for a huge explosion but unfortunately not for energy generation.

Until now, the most promising results have been obtained by using hydrogen isotopes, that is, different molecular forms of hydrogen. Principal among those have been two of the 'heavier' forms of hydrogen, deuterium and tritium. The former can be obtained from seawater and is plentiful but the latter is rather less so and has to be manufactured mainly from lithium. However the biggest problem is that a temperature in excess of 100 million DegC is required for the reaction to occur! By using special equipment and vast amounts of electric power, this reaction has been achieved experimentally and nuclear fusion power has been generated for one or two seconds. In the most successful experiment to date, the generated power was claimed to be around 60% of the input power. This is encouraging.

Late in the 20th century, there were scientific claims that nuclear fusion power could be obtained by a 'cold' process – that is, a chemical reaction taking place at normal temperature and pressure in simple laboratory equipment. This caused great excitement and many rushed to replicate the process – regrettably with no success. Unfortunately, this 'cold fusion' process is now generally discredited, though some scientists continue to work on it. In more recent years, there has been a claim that a process called 'bubble fusion' can be created simply and with little equipment.

Once again, it seems that these experiments cannot be replicated either.

At time of writing (2010), the situation is that no one has produced a working nuclear fusion generator (either 'hot' or 'cold') that can produce an excess of sustained power. This of course is the requirement for a commercial nuclear fusion generator. However great enthusiasm remains and much experimental work continues in scientific establishments around the world.

In 2006, an international body was created to conduct new experiments in nuclear fusion research. This is the ITER project which is expected to last for 30 years. The project is funded by many of the major countries in the world. The first phase of the work has produced theoretical plans for a new nuclear fusion reactor which is expected to deliver an excess of power. It is hoped that the first experimental version of this new reactor will be ready to operate in 2018.

If nuclear fusion could be made to work commercially, it is no understatement to say that it has the potential to solve the energy problems of the world. It has been suggested that a small quantity of nuclear fusion fuel (perhaps as little as 25 grams) could supply a person's total needs for electricity throughout their life.

Energy efficiency

Strategic methods to reduce energy consumption

In the 21st century, the adoption of energy efficiency has been recognised as a viable way to achieve a reduction of global energy consumption. It is reasonable to assume that energy efficiency is part of the design and construction of every new product and technology. Every designer and engineer seeks maximum efficiency from their products; those who buy and make use of them seek and expect efficiency too. Today, the energy efficiency of new products is pursued ingeniously. In the past, a lot of equipment released large amounts of wasted energy; now, this energy is carefully captured and reused if at all possible.

Obviously, the replacement of old, inefficient and polluting installations with new, more efficient and 'cleaner' equipment is another contribution towards the solution of the predicted global weather and climate problems. A coordinated approach is essential, involving global agreements, national initiatives (in concert with the scientific community) along with the cooperation of industry and, notably, every individual too.

Global agreements have already been discussed with respect to the four world dangers identified earlier (ozone depletion, acid rain, global warming, global dimming). Science has been empowered by world politicians to study and advise; the outcome has been a series of agreements designed to minimise the problems.

These strategies have been adopted in individual countries across much of the world. Business, industry and individuals are kept fully informed of the situation and encouraged to adopt the means and incentives to achieve efficiency. Transportation has been designed to reduce energy usage. Alternative means of transport (e.g. canals) could also be part of the solution.

It is noted that technology has already had some impact on business travel; for example, video conferencing may make it unnecessary for business people to travel vast distances to sit around a table for a single meeting; online computer networks mean that many people may work from a home office and avoid commuter travelling. Such philosophies are highly effective from an energy saving point of view.

It is accepted that the industry of the developed world is often efficient in its use of energy. This is not altruism. Modern industry is profit motivated and all costs are examined carefully to seek maximum efficiency. Energy will not be wasted if it can be captured and reused in some way. Regarding pollution emissions, many national laws require that these are carefully monitored and minimised as far as possible, perhaps by systems of capture.

Of more concern, however, is the industry of the developing world. Here, the best technology is not always available and industrial processes may well be similar to those of many years ago, when there was much less regard for aspects of pollution or energy use. It is in the interests of the whole world that these industries be assisted to achieve less pollution and greater energy efficiency. The morals and ethics of the Christian World would suggest that developing world countries should be assisted as far as possible, for the good of all. This is surely an action of responsibility, stewardship and love.

A focus on individuals

While it is accepted that most individuals will not have direct powers of decision at global, national or large-scale industrial levels, he or she is nevertheless a very important person in the process of energy efficiency. Essentially, it is the individual who will need to act to save energy; therefore it is the individual who needs to understand the issues involved so that their cooperation may be fully gained.

It is probably at the personal and family level that energy efficiency is at its worst. Most households are not run by accountants and while families are unlikely to waste energy grossly (like leaving the hot taps running all day) there is not usually a tight control on energy efficiency. It is suggested there are two areas where there is considerable wastage of energy at the family/individual level. These two areas concern domestic accommodation and personal transportation.

- *Domestic accommodation:* sadly, many domestic houses have been poorly constructed from the energy efficiency point of view. Also, a lack of maintenance is likely to affect the energy efficiency of a building and its equipment. In cooler climates, badly insulated houses are often heated by inefficient heating systems and a vast amount of heat energy is lost through walls, windows and roofs. Icy draughts of air enter through poorly sealed doors and windows. Traditional central heating boilers (still common in many places) pour large quantities of heat and pollution into the atmosphere all the time they are in operation. Hot water systems heat large poorly insulated tanks of water to high temperatures and then allow the heat to radiate wastefully away.

  In hot climates, the problems of construction, maintenance and usage are the same although the temperature situation is reversed. Here, the excessive heat is countered by powerful air conditioning systems which require large quantities of energy. The cooled air is lost through poorly insulated walls doors and windows. All over the world, a great deal of energy is wasted by inferior building construction and poor insulation.

Energy efficient houses are not only much cheaper to live in but a great deal more comfortable. To address this problem, governments and local authorities everywhere are encouraging and (sometimes) assisting their individual citizens to make their houses more energy efficient. New-build and rebuild properties should be built to very high energy efficiency standards; this is already happening in some countries. Furthermore, the pollution products of domestic energy production and usage should be carefully controlled.

Everyone should be encouraged to install personal energy generating devices and given advice and assistance to do so. Today there are 'power generators' that replace domestic central heating boilers; these not only provide heating for the house but generate electricity in the process. This electricity can be used by the householder and any excess sold back to their electricity supplier. Of course there are the other generating devices to generate power from the sun, wind or water. All these methods should be maximised for the personal benefit of the householder and to assist in the reduction of energy usage and pollution problems.

- *Transportation and the motor car*: when global energy consumption was discussed in Chapter 9, transportation was identified as a major energy user. It was proposed that the citizens of today's world travel further and more frequently that ever before, helping to consume vast amounts of world energy as they sit aboard ever more powerful aircraft, ships, trains and buses. The motor car was also mentioned as a vehicle capable of stirring up much passion, although the energy consumption of individual motor cars is not in the same league as the other forms of transport mentioned above.

However when the question of efficiency is introduced into the equation, the situation is altered. The fact is, the private motor car is used all over the world in an extremely inefficient way. This is not the fault of the motor car manufacturer, since modern cars have become much more energy efficient and 'green' when compared to the cars of even a decade ago. The problem here is the way the private motorist uses their car. No matter where one goes in the world, many (perhaps most) cars are seen with one individual on board – the driver. Furthermore, many of these cars are large vehicles with powerful engines.

Although public transport road vehicles have engines that use far more fuel, they offset this by carrying many passengers. A comparison can be made by looking at how many kilometres an individual may travel on various forms of surface transport for the consumption of one litre of petroleum-based fuel; studies into this usually suggest these sorts of figures:

Distance travelled per person using 1 litre of petroleum-based fuel	
Train with 300 people on board (assumed average)	55km
Bus with 40 people on board (assumed average)	50km
Small private car with 1 person on board	14km
Large private car with 1 person on board	7km

Obviously these are general figures and any increase in the number of people travelling in either of the cars increases the km figure on the right. It is a direct relationship; the more people on board, the more the 'kilometres per litre of fuel' figure increases and the more efficient that form of transport becomes. The efficiency message is clear.

While public and commercial transportation do indeed consume large quantities of energy, there are considerable economic pressures upon them to operate efficiently; in today's world, cost-driven commercial activities abound. It is clear that the same cost-driven pressures do not apply to the individual. It is certain that no-one could financially justify the purchase of a large, powerful and very expensive motor car for the transportation of a single individual. The reasons for such personal purchases are more to do with passion than efficiency!

This inefficient use of the private motor car is judged to be a serious problem because there has been a large increase of personal vehicle ownership across the world. In developing countries, increasing affluence has brought a significant increase; even in the developed countries of the West, car ownership statistics show an increase in most countries. While the world population has doubled in the last 50 years, private car ownership has increased by a factor of 10. Today, almost 50 million cars are produced each year. This explosion of car ownership contributes in a major way to the increase of greenhouse gases measured in recent decades.

What more should we be doing?

The last sections have suggested some of the practical actions that energy-using individuals could take to reduce energy and pollution. It is thought likely that many individuals would have no objection to adopting at least some of the technology, especially if it had significant personal benefits. The trouble is, setup costs are often very high and such costs are not recouped for many years. Of course the increased purchase of energy efficient technology will in time bring setup costs down but it is suggested that there would still be a need for serious supportive action from governments and local authorities; in other words, significant subsidies are required.

There is no shortage of advice about what to do. From those who broadcast warnings that human activity is making a significant and increasing contribution to weather and potential climate change, the recommendation is unequivocal. They say there is a need to take global and coordinated action to reduce energy consumption from fossil fuel burning; such action will reduce pollution and the effect will be to slow the rate of global warming and global dimming and reduce acid rain. Regarding the very real dangers of ozone depletion, the recommendations are for increased vigilance to stop the production and use of illegal CFCs, plus action to include the newer manufactured gases that cause similar dangerous effects.

There is however variation of opinion in the urgency of the actions to be taken. From those who are totally convinced that the dangers are real and imminent, the call is for immediate and drastic measures. They recommend that the pollution of fossil-based energy consumption must be cut at all costs, this to be achieved by global, national, industrial and individual action. For others the situation is still under review and so their recommendations tend to be confined to easier options, like the strengthening of current international agreements and the negotiation of new ones.

For those who do not agree that human activity is responsible for global warming, etc., or for those who do not accept the evidence produced by the majority, no requirement for energy saving is seen to be required. They are opposed to the recommendations of the others. In addition, there are billions of people worldwide who are ignorant of the issues or do not care about them; for them, the issues are of no interest and therefore no question of lifestyle alteration is likely to be countenanced, now or in the near future.

In a very real sense, the last paragraph defines the problem today. Although there can be powerful international agreements regarding global energy and national government laws to apply these agreements in each country, the essential actions to save energy devolves largely to the individual who needs (ideally) to be persuaded to comply. If persuasion had to be replaced by unpopular legislation, damaging law and order issues may well be generated with highly undesirable outcomes.

The next chapter will focus squarely on the individual to evaluate for them the personal realities which are likely to accompany the occurrence of weather and climate change as well as the relevance and significance of their personal energy saving actions. Additionally, the role of 'Christian World' morality will also be shown to be a fundamental factor in their choices and actions.

11

Are We Free To Choose?

Choice in the Christian World

The Bible is unequivocal. Man was created extremely finely by God and every single person throughout time has been given intelligence, sensitivity and the capacity to love. To this was added complete freedom of will, so complete that people have always been able to choose to reject their very creator, God himself. Mankind's creation, blessing, authority and responsibility has been comprehensively explored earlier in this book; also, the disobedience that resulted in the 'Fall from Grace' has been shown to be a significant indicator in the story of mankind.[1]

By the symbolism of 'eating the Forbidden Fruit', Adam and Eve demonstrated with certainty that they had complete and absolute freedom of will; freedom to disregard and disobey the most absolute power over them. At the same time, they showed

1 Genesis 1-3: all verses

how human judgement can be very seriously flawed. Clearly, this is nothing new. Human judgement since the beginning of Man's creation (whenever date you choose) has always featured great fluctuations. Down the centuries and millennia, people sometimes judge, speak and act with great negativity, marked out by uncaring, selfish, even brutish action and attitudes; at other times, their judgements are fine and compassionate, their actions steeped in love. The early chapters of the Bible describe how the inhabitants of that world lived in those fluctuating terms. Look around; those same terms describe today's world with remarkable accuracy.

That very same freedom of choice illustrated by the Garden of Eden story is an integral part of the law and morality of the Christian World. The Christian World is a free world, where people are allowed to make their own choices, constrained only by such laws that are necessary to make community and society work for the benefit of all. So people exercise their free will as they make choices on personal, family, community and country-wide matters. To do this, they review and study relevant information, by applying their God-given intelligence and sensitivity, further refined by God's goodness and love. Most important of all, this process applies to everybody whether they know God or not; for he knows them.

Your personal choice

'Why doesn't the Government do something about it?' We are all tempted to cry this when confronted with a problem that appears to be outside our control. This may well be what some people say about the weather and climate change issue. In this case, there is no doubt that governments are certainly trying to do something about it; Chapter 10 has shown that there are major, complex and serious initiatives in progress and some of these have been in place for decades. Of course we would expect governments and other responsible authorities to take global warming, weather and climate change seriously.

However the actions of national and international authorities are only a part of the solution to the energy and pollution problems. The other part concerns our own personal attitudes which are shaped by our judgements. Specifically, solutions to the problems concern the personal choices we need to make, because everyone who lives in a modern technological environment is a significant user of energy. Clearly, the issues concerned with energy, pollution, and climate change will involve many in the process of making choices, both collectively and individually and the outcome will be determined by these choices.

Science can provide all the facts it knows – and it has. No doubt it will continue to do so. The realities of global warming have been made clear. The expert predictions and interpretations of weather and climate change have been carefully spelled out. Details of the outcomes are rather more problematic because each predicted scenario is necessarily based on a degree of speculation. Unfortunately, the most dramatic implications have often been singled out for popular publication and this can distort people's perception, which in turn can hinder personal understanding of the overall situation.

The spread of scientific opinion raises further difficulty, although this can be regarded as inevitable in human society. Apart from a situation of absolute certainty, there are always detractors from every theory and disagreements about every assessment. This is a constant part of humanity all over the world and it certainly true in all parts of the Christian World. However this hubbub of argument and dissent further complicates matters for those who are living the energy-hungry lifestyle of the developed world. It is now appropriate to present a final review of all the issues, this time sharply focussed upon the implications for the individual.

The realities: Global warming effects and us

There is no doubt that global warming is a reality. The facts and implications impinge on each of us, personally. Those facts and implications are discussed in the paragraphs below.

Rising sea levels

- *Ice melting*: both sea and glacier ice have been melting at an increasing rate as a result of the slowly rising air and sea temperatures and this process has recently been accelerated by marked changes in surface reflection factors. An ice sheet presents a highly reflective surface to the sun's radiation, absorbing little and reflecting most of it away. This helps greatly to maintain the extreme coldness of the ice. A snow field on land (or ice) has exactly the same effect.

When the ice melts away, both land and water surfaces become considerably less reflective so more heat from the sun is absorbed. In the case of land, the sun heats a relatively shallow top layer of the ground and this inhibits any future ice or snow cover; the air above the land surface is then heated by conduction and the heat is spread upwards by convection and turbulence. By contrast, the sea presents essentially a dark but transparent surface to the sun. The heat of the sun is able to penetrate into the water and raise its temperature; the thermal circulations which are then formed spread the heat downwards, causing the sort of sea warming that has been reported by the sea temperature studies. So on land and in the sea, global warming is boosted by snow and ice melting.

The melting of floating sea ice (icebergs, ice floes, etc.) does not raise the level of the sea because no extra water is being added. However ice melting from land adds to the water volume of the sea and causes the sea level to rise. There are considerable ice field areas that have formed on top of land; much of the southern polar ice fields are in this category, as are most glaciers elsewhere in the world. Melting ice from these sources has already caused the sea level to rise. Some of the melting rates have been very rapid and are causing great concern.

There is now a further worrying factor. Until recent years, it had been assumed that ice fields would melt from their top surfaces and that the addition of ice water into the sea would be relatively slow. It has now been shown that the melting ice may not just run off into the sea but penetrate downwards through cracks and fissures that will develop through the depth of the ice. This will form pools of water at the base of the ice mass which is frozen on to the land below.

In time, the replacement of solid ice with a layer of water will cause huge volumes of ice to be detached from the land surface. If the land slopes towards the sea (a very common configuration), these huge volumes of ice, lubricated by the water, could slip suddenly down the slope into the sea. This would not only raise the level of the sea suddenly but generate huge tidal waves of tsunami proportions that are certain to cause catastrophic flooding and destruction on land areas even thousands of miles away.

– *The increase in sea temperatures*: the increase of sea water temperatures also generates sea level rises by expansion. Any temperature rise in water above 4 DegC causes it to become less dense and expand to a greater volume. Except for areas adjacent to the northern and southern polar ice fields, the world's seas are normally above 4 DegC. The greater volume of water that is the result of this expansion is added to the rising sea level caused by the ice melt.

Furthermore, the rise of sea temperatures has been shown to have a significant effect on aquatic life. Although the recorded temperature rises appear to be numerically small (0.3 DegC), aquatic life is already showing extreme sensitivity to this small change; however the acidification of the seas by carbon dioxide absorption is also a factor. Coral reefs in tropical seas have already been weakened or died (by the effect of coral bleaching) and this has greatly affected the other plant and animal life that depends on them. Furthermore, fish and other animal life in the seas are showing disturbed migratory patterns and some species of sea life are now appearing in waters never visited before.

– *The effect of rising sea levels*: the rising sea levels have already brought permanent flooding to some of the lowest coastal areas of the world and this process continues. There are many predictions of the amount of increase and the timescale of occurrence; these are based on a whole range of warming/ice melting scenarios. The past decades have seen a sea level rise of several centimetres and the prediction for the next 20 years is for a further steady increase – calculations indicate that water expansion alone could add up to 8cm and the melting of ice into the water will add to this further. Predictions for the end of this century are wide-ranging, spanning from 15 cm up to 1 metre, with catastrophic ice melts forecast at more than 5 metres. The most recent IPCC predictions suggest a sea level rise of around 0.5m is the most likely.

While the figure of 0.5m seems reassuringly modest when compared to some of the more extreme propositions, this sort of rise would flood considerable areas of low-lying ground around the world and millions of people would not only be affected by this but placed in real danger. Remembering that most coastal areas are subject to tidal flows, high tides and in particular the highest spring tides would bring a much greater risk of serious flooding to areas currently safe from this danger.

Regarding the changes in aquatic life patterns, the precise way in which this will develop is unknown but it seems certain that our world's sea life is being changed considerably. How this will impinge on the life balances of our planet is a matter of considerable speculation. However one hazard that can be noted with concern is the presence of dangerous sea life in previously safe waters. Sharks and other sea predators have been found in previously safe waters, as have the most dangerous types of jellyfish, etc.

The prediction of more extreme weather events.

Unlike the absolute reality of global warming, 'more extreme weather events' remains an IPCC prediction at this time (2010), albeit now considered 'very likely'. Some who support this claim insist that the process is already being demonstrated, with every extreme weather event identified as an example. Those who disagree argue that each severe weather event is merely a demonstration of the normal variability of weather. They point to the records of extreme weather events over the past centuries and millennia.

For instance, the designated 'worst storm' to affect the UK during the 20th century happened in 1953, when the level of global warming was less than it was later in the century. Likewise, weather records suggest that the worst hurricane to affect the mainland of the USA in the last 100 years was in 1931. The Hurricane Katrina disaster that resulted in the catastrophic flooding of New Orleans in 2005 was largely a consequence of the city's location below sea level rather than the pure intensity of the hurricane itself.

On the other hand there is statistical evidence to indicate that the frequency of tropical cyclones has increased, especially in the last decade; this increase in frequency is also claimed for other types of extreme weather all over the world. This is significant, since increased frequency of occurrence is a primary way of judging the truth of the 'more extreme weather events' claim. The evidence continues to build up.

The implications for the individual

Chapter 5 identified two categories of extreme weather types. The first category referred to extreme weather events which involved weather in one of its more violent moods. This type of extreme weather is likely to affect an area for a relatively short time, anything from a few hours to a few days. A typical example would be a severe storm tracking over an area, imposing dangerous and damaging conditions for a limited period of time before moving away. The second category of extreme weather referred to prolonged periods of particular sorts of extreme weather conditions, like weeks or months of intense heat or dense fog. In the case of this type of extreme weather, the effects come more gradually, take time to build up and then last much longer. Although the actual weather events are not violent, they are equally dangerous.

The effects of extreme weather: Short-lived events

- *Wind:* wind is the movement of the air which encases our planet. These air movements can range from light breezes to storm force winds; in addition, changes in wind speed and direction can be very sudden. Although the airflows (the winds) we experience near the surface of the Earth appear to blow horizontally, that is, parallel to the ground, in fact air moves freely in all three dimensions. So there are airflows all through the depth of the atmosphere, with components of vertical movement as well as horizontal. The fastest airflows are usually to be found in the very high atmosphere, not far from the top of the troposphere. These are the 'jet streams' that are sometimes mentioned in weather broadcasts and they are an important factor in the science of weather forecasting.

Jet streams are also of great interest to aviation. An aircraft wishes to avoid flying against a jet stream because this would greatly reduce its ground speed and extend the flight time significantly. On the other hand, aircraft are happy to fly with the jet steam provided it is not too turbulent. This increases ground speeds very significantly and reduces flight times. This is how 'record' flight times are usually achieved.

Fundamentally, the airflows of our world are driven by the sun's energy, with the Earth's differential heating constantly providing the organisation. The strongest winds near the surface of the Earth are usually found in the circulations around low pressure systems, although the very large shower clouds (for instance the huge cumulonimbus which is associated with lightning, thunder and hail) can also cause very strong and gusty winds at low levels. Because low pressure systems and cumulonimbus clouds move, the stormy weather arrives at an area, lasts for a while (sometimes quite a long while) and then moves away, taking its violent winds with it.

It is obvious that wind is capable of causing considerable damage since this is demonstrated frequently during storms in every part of the world. The damage occurs as a result of the force that wind exerts against an object; rapid variations in the wind flow (gustiness) tend to increase the destructive effect. A considerable contribution also comes from the vacuum effect created to the lee of solid objects. Although structures need to be designed and constructed with wind forces in mind, it is not uncommon for them to have parts that are far from streamlined; for instance, the roof structures of many buildings have the traditional overhang to deal with precipitation but this can greatly assist a strong wind to lift the roof off; the lee-side vacuum effect then helps to accelerate the wreckage away.

Apart from direct damage, structures (and people) can be severely damaged by debris picked up by the wind. A roof lifted off a building not only damages that structure but everything else it comes in contact with it as it tumbles downwind. Trees are often broken, uprooted or blown down by the wind and much damage is caused in this way. Most severe wind events last for a relatively short period but the damage they leave behind not only causes great inconvenience but takes a considerable time to restore.

For the individual, a higher frequency of stormy winds means that greater damage will occur more widely. Even if lives and possessions are not damaged directly by a particular event, everyone is affected indirectly. Damage may be caused to the supporting infrastructure of the community, such as public utility failures (power, water) or blocked roads, etc. Also, there will be financial implications to local community charges and insurance premium rates.

– *Tornados:* tornados are very dramatic weather events often associated with tropical storms; the largest of these extremely powerful circulations are indeed found in the Tropics. However, tornado circulations can form anywhere in particular weather situations.[1] In mid-latitudes, they are formed when extreme uplift, often associated with thunder clouds, causes strong low level spinning convergence in the air below. This develops into a characteristic funnel-shaped circulation, which extends downwards from the cloud base and often reaches the ground.

At the bottom of the funnel there is a very large decrease in air pressure, causing strong upward suction at this point. The effect is like a very large moving vacuum cleaner nozzle and that is exactly how it acts! The cone of a tornado is usually visible because its shape is marked out by all the debris it has sucked up from the land surfaces. When the tornado is over the sea, the funnel become lighter in colour because it is now filled with water taken from the sea (and sometimes with fish, too).

In all cases, tornados cause much damage. The combination of wind and suction is able to lift heavy items into the air and carry them for considerable distances. From the individual point of view, direct damage by a tornado will be severe; injury and loss of life is very common, buildings destroyed, trees uprooted, etc. In-

1 Appendix B: Tornados, see p. 215

directly, the community in general will bear the brunt of tornado damage, similar to the comments in '*Wind*' above.

The effects of extreme weather: Both short-lived and prolonged

Some events of extreme weather cause great damage in the short term but their effects can also linger on for a considerable time.

– *Rainfall and flooding:* powerful weather features such as deep low pressure systems (both tropical and non-tropical) often produce large amounts of rain. Other weather systems, such as very active and slow-moving weather fronts, can also generate intense and long-lasting rain over an area. Very large and active shower clouds can produce violent rain, too. Intense rainfall, however generated, can often be the cause of localised flooding when drainage systems cannot cope with such huge and sudden volumes of water.

However many occasions of flooding have indirect components, too. In such cases, the very heavy rain falls over a large area which includes high ground. The rainwater then drains down from high ground areas along the channels of rivers and streams to reach lower ground below. This happens very quickly if the high ground is already saturated by earlier rain. The lower ground areas, which may or may not have experienced heavy rainfall directly, is suddenly inundated by large volumes of floodwater from the high ground above.

When slopes are steep and river channels narrow, the very rapidly flowing waters from above are likely to pick up all kinds of debris (trees, boulders, etc.) and carry them downstream to act as battering rams in the turbulent waters, greatly increasing the destruction below. This kind of flooding is often very sudden and catastrophic, with injury and loss of life. If wind and rain occur together, as they do frequently, then any serious building damage by wind leaves the structure vulnerable to water damage, too.[1]

There is also another mechanism which can produce serious and unexpected flooding. When a powerful low pressure system

---

1 Appendix B: Rainfall and flooding, see p.212

is situated over the sea and develops high winds around its circulation, very large sea waves can be generated. The energy in such large waves cannot be easily dissipated and so, with suitable wind patterns, the waves may develop into a powerful storm surge which bursts upon shorelines far away, causing totally unexpected and serious flooding in the coastal communities there.

So rainfall which causes flooding is an example of short-term extreme weather. Significantly, though, the flooding element can also be categorised as a prolonged effect. Although the flood has been a sudden event which, for instance, washes away bridges, roads, etc., the damage takes a considerable time to repair and this causes great inconvenience and problems. Similarly, the flooding of a building is catastrophic. The building has been designed and constructed to cope with only dry conditions inside. When flooded, the damage is severe and can take many months, even years, to restore.

All the indirect costs mentioned in the previous paragraphs apply here also. While the individuals affected directly by damage from flooding are likely to suffer the most personal expense, infrastructure repairs to roads, bridges etc. will be an indirect charge to everyone through local taxes or insurance premiums. In addition, it is relevant to note that the drainage systems of any community are designed to cope with the 'normal' rainfall situations of the area. If there is a consistent increase of rainfall, drainage systems may need to be redesigned and reconstructed at significant cost to everybody.

– *Snow and ice:* winter weather systems in mid or high latitude regions are often associated with snow and ice. Heavy snowfall can accrue very quickly on land and reach a significant depth in only a few hours. Subsequently, compacted or temporarily melted snow turns to ice. Snow and ice considerably impedes normal life and protracted occasions of this hazard can cause considerable hardship and danger for extended periods.

There is always a great public outcry when significant snowfall affects a region that normally enjoys a relatively temperate climate. In such places, a modest 10 or 15cm of snow invariably brings great disruption. The inconvenienced public point furiously to countries where the climate is much colder and demand to know why these places are unperturbed by 50cm or even 100cm of snow. The answer is that their routine infrastructure is designed to deal quickly with deep snowfall and so normal conditions can be restored quickly. Likewise, icy conditions are dealt with by 'gritting' roads and pavements quickly and as often as required. The grit normally contains salt which helps snow and ice to melt by reducing the freezing point of water.

It is clear that every aspect of the infrastructure of a region will be geared to its 'normal' climate. In temperate countries, snow events may occasionally produce 5, 10 or even 15 cm of snow and this depth of snow can (hopefully) be dealt with by a small fleet of light snow ploughs. In such temperate climates, it is unsurprising that there would be no snow clearing facilities to deal with 50 or 100 cm of snow; such snow depths require heavyweight snow blowers. However if the temperate regions began to be affected routinely by deep snow, then a new infrastructure will be required, involving considerable costs.

Snow and ice is another example of extreme weather which can be both short-term and prolonged. After snowfall, if the weather turns mild, a thaw will occur and the snow and ice will disappear. However if the weather remains very cold, even if there is no more snow, the existing snow cover will persist, often becoming more icy as partial thaw and re-freezing takes place. Then, the degree of prolongation depends upon the efficiency of the snow clearing and gritting operations of the local authorities.

Snow and ice, whether short-term or prolonged, is a very great hazard to the public. These conditions are the direct cause of many accidents – everything from a twisted ankle to a major traffic pile-up! Considerable suffering and damage is caused by these accidents and the costs recovered through taxation and insurance premiums are inevitable.

– *Frost and low temperatures:* one of the physical characteristics of water is that its frozen form is less dense than its liquid form. This is why ice floats on the surface of water. It is also why water freezes from the surface downwards, which enables aquatic life to continue in the water below an ice sheet. Were it not for this characteristic, much of the life we now know could not exist.

The lower density of ice means that water expands when it freezes and it is this expansion that causes damage to any rigid object penetrated by water. This is one reason why cracks develop in roads surfaces and likewise, in buildings and other structures into which water has penetrated. In severe cases, the road surface can be weakened so much that it collapses into deep potholes. Other structures (like buildings, bridges) can be so weakened by cracking that they eventually fail and collapse. Damage may also occur in water-filled objects such as pipes or tanks; here, the ice exerts so much pressure that the pipe or tank is split.

If directly affected by such damage, the individual will suffer considerable inconvenience and perhaps danger, too. The repairs of frost damage are invariably costly. Also, the fracture of a tank or pipe may cause flooding if the damage is within a building, with all the misery that imposes. An increase in the incidence of these events will increase costs for all through community and insurance charges.

Frost certainly suffers from the short-term damage implications stated above. It can also become a prolonged problem if low temperatures persist for an extended period. This is where physiological problems become serious. Consistent low temperatures should not be problem if you can keep warm and you are in good general health. This involves living in a comfortable well-heated environment and possessing the necessary warm clothing to protect yourself when you are outside. Unfortunately, even in affluent communities, many people do not live in such comfort and cannot afford the necessary degree of warmth; likewise their clothing is inadequate. In addition they may have poor heath and be vulnerable in other ways.

Of course the poor and vulnerable need to be supported and the public services of the Christian World should be adequate to ensure that the support is timely and appropriate. This, however, is not something that we should ignore and leave to the public services. We all have responsibilities for the weak and vulnerable and should provide help and support in any way we can. This is rightly fixed in the ethos of our christian world.

– *Dense fog and air pollution:* air always contains water in the form of invisible water vapour gas. The water becomes visible as cloud in the sky or fog on the ground. The formation of fog often occurs when the weather is relatively calm and settled. In this case there is usually little cloud and night-time temperatures decrease quickly. The cold ground cools the air by contact until it reaches saturation point. At this time, the water vapour in the air condenses into small water droplets and fog is formed. However it is also possible for fog to form in less settled weather conditions; this happens when moist mild air is blown across much colder ground and its temperature cooled below saturation point.

Fog, however thin, is always inconvenient, especially for the traveller. Here it restricts visibility and can be dangerous. On the other hand, persistent dense fog is very unpleasant. At its thickest, visibility is reduced to a few metres and road travel becomes very dangerous, especially at night. Even walking a short distance in the darkness in dense fog can be very hazardous and this is the cause of many accidents. Winter conditions of this type are even worse, because the fog may become 'freezing fog', depositing rime ice on all surfaces. Roads and walkways become even more dangerous as a result.

Many occasions of fog are temporary, characteristically forming at night and clearing away during the day. In such cases the fog becomes a short term event. However when fog becomes persistent for days or more, the situation becomes much more serious. If the season is winter, freezing conditions can persists for days or even weeks, because the fog prevents the normal daytime temperature rises. Persistent freezing fog makes moving around on foot or by most forms of transport very hazardous.

Polluted air is an added hazard. Persistent high pressure prevents low-level air from rising more than a few thousand feet from the ground and the low-level airmass becomes progressively polluted by a whole range of chemical agents emitted from vehicles, chimneys and industrial outflows everywhere. This pollution is injurious to health and both plants and animals suffer. People susceptible to respiratory and related problems often require medical help and deaths may occur.

Fog accidents or illnesses are costly to health, life and personal economy and will always impose considerable community costs through tax and insurances.

The effects of extreme weather which will normally be prolonged

– *Extreme Heat:* everything that is constructed or manufactured is designed to withstand a range of temperatures, normally with good safety margins built in. If these tolerances are exceeded, then damage of some sort is likely, often related to excessive expansion. Also, excessive heat can affect the ground environment which supports whatever is built upon it. Here, humidity may well play a part also; as the temperature rises, the absolute humidity falls because of increased evaporation. The supporting ground may then be subject to shrinkage and building foundations will be affected adversely, causing the common problems of subsidence or heave. These problems are usually very difficult and expensive to correct.

Heat damage applies not only to structures but crucially to the physiology of animals, including humans. Greater heat and less humidity mean that a greater consumption of water is required to maintain the body within its narrow tolerances for life. If the body is healthy and the availability of water sufficient, this is not normally a problem. However a weak or diseased body has a much reduced capacity to withstand excessive heat even when water is available – as the reported death toll caused by the extremely hot European summer of 2003 has shown. Almost 40,000 people died, with the highest death rates in Italy and France. Also, there was a serious economic cost when crop yields across the region were greatly reduced.

Clearly, there are serious health and economical problems associated with an extended period of heat. Costs will be applied to the individual through taxes and insurance premiums.

– *Drought:* drought describes a situation of extreme dryness over an area where there has been little or no rain for an extended period. Almost invariably, it also means that water supplies dwindle. Drought conditions are often accompanied by the excessively hot weather discussed in the previous section. In temperate (normally mid-latitude) countries, drought can be a serious inconvenience because it disrupts normal water supplies; water usage is curtailed and, in severe cases, the mains water supplies fail and water needs to be obtained from common standpipes or tanker vehicles.

The situation is much more serious in hot climates. Here, the occurrence of rain in the appropriate season is very much an essential of life. When drought occurs, the makings of a disaster soon appear. Agriculture of all types suffers. Crops fail and, unless water can be supplied from elsewhere, farm animals suffer and die. The livelihood of farmers is ruined. Other plant life and wild animals suffer similarly. Desertification takes place. In these areas, drought at its most severe causes much human suffering and many deaths.

An increased frequency and intensity of heat and drought accelerates all the social and economic problems described in this and the last section. For those directly affected, the situation may well be involved with life and death; certainly there will be economic and financial implications for the region and ultimately the world. Such expenses eventually devolve to the individual through taxation.

- *Sunshine:* absorption of the sun's electromagnetic radiation is an important factor for the Earth's plant and animal life. For humans, sunlight is our main source of Vitamin D production. Our bodies require sufficient Vitamin D to be present so that calcium may be metabolised and our bone structures kept strong and healthy. It is the action of sunlight on our skin, specifically the UVB component, that produces most of the Vitamin D that we need.

  It is possible that a greater frequency of storm and precipitation events will increase the amount of cloud present in the sky and so reduce the amount of sunshine. Where this occurs, this has the potential to cause detrimental effects on plant and animal life, including the potential for Vitamin D deficit in humans.

The Christian stewardship implications of 'more extreme weather events'

All the paragraphs above describe the personal implications of 'more extreme weather events' for the people of the Christian World in terms of inconvenience, health and financial economy.

There is, however, the Christian dimension of stewardship to consider, too. It will be recalled that this is referred to many times in the Bible, in both Old and New Testaments. The stewardship requirements towards other people in the world are stated succinctly in the New Commandment of Jesus when he directs Christians with the all-pervading commandment 'Love one another'.[1] The morality of the Christian World translates that into concepts of compassion and willingness to help, especially applied to those who are weak and vulnerable.

All the types of extreme weather discussed in previous paragraphs refer to people being placed in danger, fear and despair. Where the weather event has been responsible for injury there will be pain, suffering and, at worst, bereavement for some people to cope with. Where poverty, weakness or illness has been exacerbated by severe weather there will worry, concern and despair. In the christian world, people are sensitive to the problems of others and it is commonplace for them to 'rally round' and provide whatever help and support they are able to offer.

Furthermore, where the extreme weather has affected people remotely, that same compassion and desire to help is shown so clearly in generous donations of financial help; for some people, the desire to help is so strong that they choose to travel to the devastated area and offer themselves physically to the relief efforts. All this help, financial or more, is given to people completely unknown to the givers. It is absolutely clear that these actions are completely in accordance with the commandment of Jesus 'Love one another' and it is an important redeeming feature of the christian world.

The prediction of climate change

In Chapter 7, we saw how the same pollution effects that cause global warming have the potential to alter sea currents, affect periodic oscillations and cause the climate of certain parts of the world to be changed quickly and radically. For instance, if the

---

1 John 13:34

transport of warm sea currents to northern latitudes was reduced or disrupted, a colder climate would be imposed upon regions that today benefit from the warm sea currents; for example, western Europe. Once the sea currents change, it would be extremely difficult, perhaps impossible, to return them to the former patterns.

The IPCC warnings recognise catastrophic climate change as a valid danger but the scientific experiments that have been carried out in recent years do not indicate a significant immediate danger. Catastrophic climate change is thought to be 'unlikely' during this century. The mechanisms of catastrophic climate change will certainly continue to be monitored carefully over the next decades and no doubt there will be further refinements in this particular warning.

- *El Niño events:* the statistics of *El Niño* events indicate that global warming is affecting the long-standing sea temperature disruptions that cause these anomalies such that they have become more frequent, stronger and much more extensive in their effects. So in this sense, the climate of the areas affected has already been changed.

The personal cost of sudden climate change

Catastrophic climate change means that the 'normal' range of weather that a region has experienced for a considerable time will be altered, quite suddenly and significantly. When this happens, it means that at least some of the implications listed under 'More extreme weather events' (above) will become permanent features of the area. This will cause much disruption to normal life, because, as explained previously, every aspect of community infrastructure will have been designed to support the weather ranges of the previous 'normal' climate.

For instance, a much colder climate will bring serious problems of survival to the weak and elderly and medical facilities will need to be boosted. All sorts of infrastructure changes will be needed for buildings, transportation and utilities design; new arrangements and equipment will be required for snow and ice clearing. People will need to adapt to this new colder climate. Equipment will need to be replaced or modified to withstand colder temperatures.

Plant and animal life will suffer and change; farming and agriculture in general will need to be altered. All these changes will be achieved with considerable difficulty and at great cost, not only in financial terms but it terms of loss of life and general productivity. There is likely to be a considerable degree of chaos as these changes are made.

Likewise, the imposition of a harsher, more extreme climate will bring problems of similar seriousness. Buildings and transport arrangements will require modification or redesign to deal with the prolonged heat of the summer seasons, with implications of more powerful cooling and insulation systems. Increased water supplies would need to be provided. The diseases of warmer climates will be introduced along with the risk of more dangerous animals and insects. During the cold and stormy winter seasons, the heavier and more frequent rainfall will bring the requirement for improved drainage. People will need to adapt to the new and more extreme seasonal weather. Farming and agriculture will need to change.

The other pollutants

- *Global dimming* is caused by large particle pollution and, since it causes cooling, it acts against global warming. However it has been shown to have serious implications for regional climate change, especially in tropical regions, where is has been observed to be especially severe. Since many of the inhabitants of the Tropics depend upon reliable climate patterns to support their generally fragile lifestyles, this has had devastating effects in these regions with many occasions of loss of livelihood and, sometimes, life itself.

  Action has been taken to reduce this type of pollution which is produced by industrial plants and some vehicles with petrochemical engines. These actions appear to have had a positive effect on the problem. Because the reduction of large particle pollution comes mainly from industry, the individual's responsibility is confined to ensuring that the equipment they buy and use is 'clean', well designed and properly maintained; this applies especially to motor vehicles.

- *Acid rain:* the danger of acid rain has long been recognised and there have been international and regional actions to decrease the chemical pollution which is its cause. This has improved the situation in some formerly polluted areas. On the other hand, industrial expansion in the developing world had added new sources of pollution with areas formerly unaffected by acid rain damage now experiencing the hazard.

  Like the global dimming hazard, individuals need to be aware of their responsibility to use non-polluting equipment and machinery as far as possible and ensure it is in good order.

- *Ozone destruction:* certain man-made chemicals (CFCs etc.) have greatly affected the high atmosphere ozone concentrations that protect us from damaging UV radiation. Science seems mostly agreed on this point and considerable progress has been made to address the danger. Today, some welcome restorative results are reported. However, there is still a significant deficit of ozone, especially over the South Pole region, when existing concentrations are compared with previous norms. Since ozone is a powerful greenhouse gas, restoration of its concentrations implies an increased contribution to the greenhouse effect; however this has been calculated to be small.

  So it seems that the warning has been heeded and the situation is improving slowly. Of course there has already been an effect on human health, especially in those areas most affected by the effects of the ozone hole – the southernmost counties of the southern hemisphere. Happily, the re-establishment of more normal ozone levels should reduce the danger in the future. This may be taken as a good example of international cooperation. However, continued vigilance will be required, not only for the production and use of CFCs but for new substances whose usage will cause ozone destruction once again.

Forward projections for global weather and climate change

Forward projections are rarely totally accurate because they of necessity involve assumptions. Today, it is common for the scientific organisations involved with the issues surrounding weather and climate change to publish a range of forward projections, each based on a different global warming scenario. Inevitably, the range of projections from each organisation or scientific authority does not agree – although there are often similarities. Most scientific organisations accept that the greater the warming, the greater the effects they are likely to produce. This also impinges on timescale, with the increased warming scenarios producing the earliest changes. These scientific organisations agree with the IPCC that the current weather and climate warnings should be considered real, important and pressing.

What about my climate?

Obviously, people are very interested to know how global warming will change their specific climate, if at all. This is a complex question for two reasons. Firstly, the local climate of an area is determined on many different atmospheric scales so it is very difficult to predict what will happen in detail when the largest scale element is the driver for the change. Secondly, no one knows what degree of global warming there is going to be and in what timescale it will occur; the degree of warming will have a huge impact on climatic outcomes.

However the IPCC have attempted to advise on the impact of climate change brought about by various levels of global warming and have related this to various regions of the world. In all cases, it is assumed that the impacts would be progressively more severe with larger global warming temperature increases. The predictions have concentrated on water availability, ecosystem changes, food production, coastal flooding and impact on human health. The following paragraphs summarise the predictions by region, presented alphabetically.

– *Africa:* by 2020, hundreds of millions of people in Africa will be exposed to increased water shortages. Less rain is likely to reduce crop yields, with subsequent food shortages. Both factors would have an impact on general health, especially in diseases connected with malnutrition. Later in the century, low lying coastal regions will be flooded by the rising sea level. This will displace large populations with many social and economic problems. Cost estimates are likely to be in the order of 5-10% of GDP.

- *Asia:* by mid-century, Central, South, East and Southeast areas of Asia are likely to have less fresh water reserves as river flows decrease. The rise of the sea level will result in the flooding of coastal areas; this will be especially severe in the large delta areas where there is a high density of population. Deterioration of population health is likely in the East, South and Southeast of Asia, linked to flood and drought conditions.
- *Australia and New Zealand*: by 2020, significant loss of plant and animal life is likely in the Queensland Wet Tropics and Great Barrier Reef areas. Problems of increasing water availability will affect Southern and Eastern Australia and North Island, New Zealand. The incidence of fire and drought is likely to decrease agriculture and forestry operations in Southern and Eastern Australia and parts of Eastern New Zealand. By mid-century, flooding caused by the rising sea level and an increase of storm activity will limit coastal development and population growth in some areas of both countries.
- *Latin America*: by mid-century, hotter and drier conditions is likely to cause areas of tropical forest to be replaced by savannah. Many species will be lost. Major crops and livestock production will decline with food shortages developing. However soybean yields in temperate zones are likely to increase. Water availability is expected to decrease as rainfall patterns change and glaciers disappear.
- *North America*: warming in the western mountains is likely to cause reduced snow depth, more winter flooding and reduced water flows in the summer. Water availability in the summer will be adversely affected. However the first part of the century may see some increased crop yields. Heat waves are likely to become more frequent and last longer with negative impacts on health in general.
- *Polar Regions*: continuing sea and glacier ice melting will change many ecosystems. There will be negative effects on many organisms, including birds and mammals. Some species are likely to be lost. The changing snow and ice conditions will

affect the human communities variably. Traditional lifestyles are unlikely to be able to continue.

- *Small Islands*: rising sea levels and storm or storm surges will result in flooding or inundation. Island communities will be displaced or driven totally from their island homes. Coastal conditions (e.g. beaches, corals) are expected to deteriorate. By mid-century, less rainfall is likely to affect local water resources seriously. The increased invasion of non-native species is expected to occur, especially around mid and high latitude small islands.

Climate change conclusions

At time of writing (2010), there is still no absolute certainty of outcome, whatever actions are taken; such is the complexity of worldwide weather and climate processes. Clearly, statistical evidence and continuing research will help to build more certainty – and this has already happened to some degree. The exception in all this is the undoubted rise of sea levels; this has occurred, is occurring and will continue to occur. The extent of sea level rises will depend, at least in part, upon the actions of the world. We should never forget that the 'actions of the world' are the cumulation of individual actions everywhere, including ours.

Are we free to choose?

Clearly, we are. However it is essential that we make those choices in the knowledge of all relevant factors. The next chapter will show how the issues around global warming, weather and climate change extend well beyond purely technical considerations.

12

The Balance Of Life

Why should I bother?

A common question, succinctly put! Despite its suggestion of truculence, this question is a totally reasonable and almost inevitable human response to all the information given in the previous eleven chapters. In considering any course of action, people want to examine all the alternatives before deciding what to do – and 'no action' can sometimes be the right decision! Furthermore, this is yet another example of the human intelligence and free will mentioned previously a number of times.

In the case of global warming and climate change, the individual's argument for 'no action' is likely to look something like this:

> 'Why should I bother to be energy efficient when there is hugely increasing energy usage in the rapidly developing industries of Asia and Africa? Why should I bother to be less polluting when those same sources are releasing ever-greater amounts of pollution into the atmosphere? In any event, my best efforts to reduce energy and pollution would be totally

insignificant - and I think my personal 'carbon footprint' is already very small.'

It is true that the mitigating efforts of a single individual or family must be completely swamped by large-scale industrial energy use, although the cumulation of all individual efforts across the Christian World would certainly be considerable. This, however, does not answer satisfactorily the pointed personal questions posed in the last paragraph. Although previous chapters have examined many of the issues connected with global warming and climate change, as well as their link to energy usage, there are still some significant matters outstanding; these matters are powerful factors in the personal decision-making processes now to be focussed upon here.

The scientific aspects

These have been examined comprehensively throughout the book and, from a scientific point of view, the situation is clear. In summary, global warming is a certainty and its effect on the seas has been observed; it seems certain that the progress of the sea level rise will depend upon the extent of future global warming. In addition, more extreme weather events are now predicted to be 'very likely'. Some rapid climate changes have been observed (e.g. *El Niño* events) but major sea current alteration is considered unlikely for the time being.

The need for urgent global action has been accepted by many countries. Global actions are taking place at the highest scientific and political levels and much discussion continues. However, it is obvious that the actions of individuals worldwide will also be of fundamental importance; essentially, it is the cumulation of all their actions that will define what happens. This makes it clear that global warming has now become personally relevant for everyone.

...but what if weather and climate change doesn't happen?

Against the torrent of recommendations for action, people remember the objections of the dissenters and wonder:

> 'At the moment the warnings of weather and climate change are predictions – and predictions can go wrong. What if weather and climate change doesn't happen? Are we making all these changes for nothing?'

As with the earlier personal questions, it is completely reasonable that people should voice these questions.

The answer is:

> 'It is correct that the weather and climate change warning are predictions. But, in part at least, these are predictions that are thought to be 'very likely' to happen (in the case of 'more extreme weather events'). However, if the predictions were wrong and weather and climate change did not happen, there would still be seriously negative issues connected with energy usage - this will be discussed in the section "Economic aspects" below; meanwhile, the reality of global warming will continue and the severity of its effects will be determined by the strength of mitigating actions taken worldwide. Therefore the scientific case alone is a powerful reason for personal action to help curb global warming'.

Economic aspects

The changing patterns of energy usage

Global energy use is increasing dramatically. China and India each have a population of over one billion and both these countries are expanding industrially at a very rapid rate. China, with 1.4 billion people, is the prime example of what is happening. To meet its industrial expansion, it has had to develop its energy-producing means significantly. This has meant commissioning many new coal-fired power stations and burning some of its large reserves of coal.

In addition, China now needs vastly more oil and gas for its industrial expansion. This huge country has now become the second largest user of oil products in the world and, because of this, has become a major importer of oil. India is not far behind in the expansion of its energy requirements. Furthermore, other parts of Asia and Africa are also engaged in similar rapid development. Construction projects in all these countries require large increases in building materials like cement, steel and wood and worldwide shortages of construction materials have already been observed.

There is also a marked effect on the individual affluence of local populations in the industrial expansion areas. As the new industries develop and start production, workers need to be up-skilled to deal with new and more complex equipment. So the wages of the workers increase and standards of living rise. Naturally, the workers quite rightly aspire to the lifestyle that the developed world has had for a considerable time, especially in terms of housing, domestic equipment and motor cars. All these changes result in greater commercial, domestic and personal energy usage and this boosts the global energy requirement even further.

Unlike the weather and climate change issue, which still contains uncertainty, the forward projection of global energy usage can be foreseen with more clarity. The developing world will need vastly more energy to fuel its expansion. New energy generation will meet some of the requirement at considerable cost; new power stations are expensive. If the new energy generation cannot meet all of the increased demand (which is the case at present), global energy shortages will boost costs still further; it should be remembered that a large component of energy price levels resides in 'supply and demand' concepts.

World population predictions are also part of the scenario. Today's population of almost 6.8 billion is predicted to rise to over 9 billion by 2050. This is not expected to be an equal growth across the world. Most studies of population place the increases in the 'less-developed' countries, while the populations of the developed world are not predicted to rise. Of course, most of the populations in less-developed countries are generally low energy users at present but, in many places, this is likely to change markedly in the future.

The increasing population figures in developing countries suggest that even bigger industrial expansion will take place in the coming decades. This will bring increasing affluence and technology to their populations; this is of course a welcome development for humanity. However it will make these populations individually hungry for world energy in a way they are not now.

Most forecasts of future energy requirements suggest at least a 50% increase within the next three decades, mostly required by the developing world.

The consequences for the developed world

The consequences for the high energy users of the developed world can therefore be seen as a progressive scarcity of energy and materials, as well as significant increases in their purchase cost. At a personal level, this translates into a worsening lifestyle, especially if no action is taken to reduce personal energy consumption. Even if individuals are willing to meet much-increased energy costs, it is possible that the supply of energy may not be so available or as reliable as it is at present. Thus, personal action to save energy not only helps to check global warming but contributes meaningfully towards the maintenance of individual lifestyles and current standards of living in the developed world.

Global interdependence

It's one world

Several centuries ago, the people of the world had little knowledge of other communities, even in their own country. Relatively few people travelled away from their home area. Transportation was slow and difficult. Communication with nearby communities was minimal and contact with more remote communities non-existent. As the centuries passed, travel eventually became easier and science provided the means of remote communication. It then become possible for people to gain knowledge about other communities, countries and continents. Even so, the people of the world continued to be generally insular, thinking in terms of self, family and their immediate community.

Nowadays, awesome developments in science and technology have made the world a small place. A great many people have access to information about the whole world; news from the other side of the world is now available in 'real time' – as the events happen. Importantly, in the last decades, it has become increasingly obvious that the peoples of the world are completely inter-dependent. Although as individual human beings we have retained a 'village mentality', we have in fact become a global village!

There is no doubt that this has brought a new understanding of the individual's place in the world. Knowledge of the world and its events have shown people everywhere that it is in their own interest to consider others, even if they are far away, and to cooperate with them to solve their problems for the benefit of all. The 2007-9 world economic recession has shown all people that global problems are of primary concern to everyone and that such problems can only be solved by coordinated global solutions. It is hoped that the coordinated and increasingly unified responses seen in the economic crisis will encourage even more responsible attitudes towards global warming and climate change.

Morality and ethics

The 'Christian World' dimension

Chapter 1 of this book proposed that the Bible and its contents were a mainstay of the Christian World, which covers a very large and significant part of the Earth and currently includes over one-third of the global population. The Bible, specifically the teachings of Jesus Christ in the New Testament, became the basis for law in the Christian World and this remains the case today.

However, the biblical influence goes deeper than mere laws, because it also determines the morality and ethics of the people who live in the Christian World. It is for this reason that Christian World people are committed to fairness, justice and compassion. These are precisely the qualities that Jesus represented and he preached them constantly to his followers. He also emphasised the wonderful capacity of every person to love, a gift that can be seen clearly in so many people.

While it is equally true that people are capable of being selfish and unkind at times, there is nevertheless a constant pressure within each person to be kind, compassionate, loving, generous and helpful. For the Christian, this pressure is focussed by the teachings of Jesus, clearly and succinctly expressed in his New Commandment 'Love one another as I have loved you...'[1]However, there is no doubt that this way of life is also embraced by the people of the Christian World, as they act with their ethos of justice and compassion.

Although extreme selfishness is rightly judged to be negative, it is certain that every human being must be self-focussed to some degree, in order to protect their life in the widest sense. This self-focus will certainly extend to those closest to them, firstly their family, then friends, neighbours, etc. Thereafter, the love and goodwill which they will extend to their community, their country and then out to the rest of the world will normally be subject to a decrease in intensity with distance; however this does not mean that their love and goodwill is non-existent, however far away the target.

This love and goodwill is so clearly illustrated when the people respond to pleas for help from countries far away, for instance places where natural or man-made disasters have occurred. The response invariably produces vast amounts of money, food and other goods in generous donations. The same generosity can be seen in large charity appeals which frequently raise currency millions. All this is a clear example of the love and compassion that is part of being human.

Stewardship and responsibility; the fundamentals

In an ethical sense, it has long been realised that the affluence of developed countries and their access to cheap energy has been at the expense of the people in the developing world. At the beginning of the 21st century it is totally unacceptable that many people still live in primitive and deprived conditions, with no access to the most basic of life necessities – clean water, reliable food supplies, proper shelter and effective medical treatment.

The industrial developments now taking place in the developing world will undoubtedly improve the lot of many people there and give them access to what the people of the developed world have taken for granted for a considerable time. The fact that

---

1 John 13:34

the poor and deprived populations of the world will become more affluent and require more world energy is something to be welcomed and celebrated.

Finally, it is no accident that all the major administrations, scientific organisations and religions of the world strongly recommend good stewardship of our planet. It is a sad fact that the actions of mankind, especially in more recent centuries, have been uncaring of our planet. Its resources have been plundered with little or no thought for its regeneration while its fabric has been polluted in many ways. This has become increasingly obvious as the population of the world increased rapidly. However it is significant that the same moral responsibility is strongly urged in ancient texts written many thousands of years ago, proving that the ethical dimension in human civilisation is far from new.

'Now Choose Life'

The words above come from the book of Deuteronomy, in the Old Testament of the Bible. Most of this book consists of a comprehensive range of instructions from God to the Israelites, communicated through Moses. Having completed many pages of mandatory instructions, God then invites the people to make a choice between 'life' or 'death'; [1] the promise is that obedience to God will bring life and prosperity, while disobedience will result in death and destruction. God's final recommendation is: 'Now choose life.' [2] This would seem to be sound advice, indeed!

In the New Testament of the Bible, Jesus taught many times about living a life of kindness, generosity and compassion. He emphasised that such actions are the only way to achieve true happiness and fulfilment on Earth; furthermore, he taught that christian action and belief are the guarantee of eternal life. This promise is expressed succinctly in John's Gospel, in what is arguably the best known verse in the Bible: 'For God so loved the world that he

1 Deuteronomy 30:15
2 Deuteronomy 30:19

gave his one and only Son that whoever believes in him shall not perish but have eternal life'.[1]

This is Christianity with a personal and individual focus. It is this focus which leads directly to the standards of compassion, justice and fairness which is the mark of the Christian World. Each one of us has been given free will - completely free will. Therefore we are free to choose and free to act. This book has set out the case for taking direct and personal action to save energy and reduce pollution. The case has been argued through science, economics, sociology and today's global realities, while the morality and ethics of the Christian World have reminded us that we are bound to act with fairness, justice and compassion.

Crucially, we have seen that intelligent self-interest and all-encompassing love for the world are not characteristics set far apart as the extremities of the human spectrum; they are in fact totally compatible and they define with accuracy the human condition in the Christian World. Both of these characteristics embrace thoughtful and meaningful stewardship of the world and all the people and other life who live upon it, no matter where.

So be wise. Choose life. The life you have now. The life you will leave for your descendants.

---

1 John 3:16

Reference Section

Appendix A: Biblical References

Bible Ref.	Subject	Chap.	Page
Genesis 1-2:all	The Creation	2	27
Genesis 1-3:all	Creation and Fall	11	173
Genesis 1:11	Plants, vegetation	2	34
Genesis 1:26	Dominion over creatures	10	153
Genesis 1:26-29	Responsibility	4	65
Genesis 1:27	Man in God's image	9	150
Genesis 1:28	Subdue the Earth	10	152
Genesis 1:29	Food for mankind	9	141
Genesis 1:30	Food for animals	9	141

Bible Ref.	Subject	Chap.	Page
1Sam. 12:16-18	God's weather control	6	101
2Samuel 1:21	Bad weather punishment	4	74
2 Kings 2:19-22	Water pollution	2	30
Nehemiah 9:12	God in the cloud	6	102
Psalm 78:14	God in the cloud`	6	102
Psalm 89:14	Righteousness and justice	9	150
Prov. 15:26-28	Obedience and dissent	3	57
Isaiah 5:6	Disobedience and drought	4	74
Isaiah 5:30	Darkened sun	4	72
Isaiah 13:10	Darkened sun	4	72
Isaiah 24:4-17	Drought and heat	3	44
Isaiah 51:6	Great heat	3	44
Ezekiel 38:17-23	God's weather control	6	102
Joel 3:15	Darkened sun	4	72
Amos 4:7	Drought	7	115
Micah 4:4	Tree benefits	2	34
Haggai 1:1-11	God's drought	6	102
Zechariah 10:1	Seasonal showers, clouds	4, 5	74, 89
Zechariah 14:17	Drought	7	115
Matthew 5-7:all	The Sermon on the Mount	1	21
Matthew 5:3-10	The Beatitudes	1	21
Matthew 5:43-48	Love your neighbour	1	22
Matthew 6:9-13	The Lord's Prayer	1	22
Matthew 16:2	The red skies	5	89
Matthew 17:5	The Transfiguration	6	103
Matt. 22:36-39	Love your enemies	1	21
Matthew 24:29	Darkened sun	4	72
Matt. 24:29-41	The End Time	3	45
Matthew 27:45	Darkened sun	4	73
Mark 4:35-41	Jesus calming the storm	6	103
Mark 4:39	Jesus calming the storm	5	88
Mark 6:45-51	Jesus walking on the water	6	103

Bible Ref.	Subject	Chap.	Page
Mark 12:29-31	Love your neighbour	1	21
Mark 13:24	Darkened sun	4	72
Mark 14:12-17	Jesus and food	9	142
Mark 15:33	Darkened sun	4	73
Luke 5:29	Jesus and food	9	142
Luke 6:27	Love your neighbour	1	22
Luke 10:27	Love your neighbour	4	66
Luke 11:2-4	The Lord's Prayer	1	22
Luke 21:27	Jesus coming in a cloud	6	103
Luke 23:44	Darkened sun	4	73
Luke 24:42-43	Jesus and food	9	142
John 3:16	God so loved the world	12	204
John 6:38	Obedience	3	57
John 13:34	Love one another	1,4,10, 11,12	22, 73, 154, 189, 201
Acts 3:19	Fallibility and weather	4	67
Acts 14:17	Seasonal weather	4	74
1Cor. 1:10-13	Dissent	3	57
2Peter 3:10	Extreme heat	3	45
Revelation 2:7	The Tree of Life	2	35
Revelation 8:5-7	The Earth burnt up	3	45
Revelation 9:2	Power of evil	4	73
Revelation 11:18	Time for rewarding	4	66
Revelation 14:14	Jesus coming on a cloud	6	103
Revelation 16:8-9	The Earth burnt up	3	45
Revelation 16:18	The Earth burnt up	3	45

Appendix B: Weather Explanations

Climate and weather

Wherever you are on Earth, there will be day to day variability in the weather. Each 24 hours presents a succession of weather events – temperature changes, degree of cloudiness, rain starting and stopping, etc. The 'climate' of an area or region is a cumulation of all the weather that has occurred there over an extended period. The word climate is derived from the Greek word '*klima*' which means 'inclined', referring to the effect of different inclinations of the sun.

From the time that the Earth was recognised as a large sphere, it was known that the hottest areas would be those places whose land surface was presented most squarely to the sun's rays, that is, the equatorial regions. To the north and south of the Equator, the Earth's surfaces are presented more and more obliquely to the sun, so the its heat has to be spread over progressively larger surface areas. Near the Poles, surfaces are presented very obliquely so the heat from the sun has to cover a vast areas of land, water or ice and heating values per unit area are very low.

Of course many factors in addition to sun rays come into play to define the specific climate of an area. The sea will have a moderating effect on islands and coasts while mountains attract cloud and precipitation. If a region tends to be cloudy, temperatures are expected to be mild during both day and night. Conversely, clear skies lead to harsher temperature variability – deserts are typically burning hot by day and very cold at night.

Rain and showers

Meteorologists always differentiate between rain and showers and the public perception is that showers are precipitation events that are of relatively short duration while rain tends to be rather more persistent. This perception is quite correct in most cases. However, meteorology defines precipitation as 'rain' or 'showers' in accordance with the mechanism that produces it.

Showers

Showers are produced by the action of convection currents in the atmosphere. The domestic convection heater (so-called because its heat is spread around by convection currents) illustrates how these currents work. In a room, heated air rises vertically from the top of the heater, flows across the ceiling and then sinks back down towards the floor, cooling as it does so. The cool air then flows across the floor to re-enter the heater at the bottom. So a rotating current is formed and maintained.

For convection currents to form in the atmosphere, it must be in an 'unstable' condition; this means that the rising air remains buoyant and so continues to rise spontaneously. So the air is heated by hot ground, becomes less dense and rises. The air will continue to rise as long as it is less dense (warmer) than the environment. At some point, it will stop rising because it has cooled to the environment temperature and is no longer buoyant. Then it will flow horizontally for a short distance until, cooling further, it sinks back towards the ground, to be drawn towards the spot where it started rising. Thus the convection current circulation is established.

Air always contains moisture in the form of invisible water vapour. When the air rises, it cools and may eventually reach its condensation level, at which point the water vapour becomes visible water droplets. At this level, white and puffy cumulus-type clouds will be formed. When these cumulus clouds become sufficiently developed, showers may be produced from them. The single word 'showers' usually means 'rain showers'; sleet, snow or hail showers usually have the preceding descriptive word added.

Once formed, shower clouds tend to move with the prevailing wind, treating the ground areas below them to periods of sunshine and showers. When atmospheric convection currents are strong, the showers may become very heavy. The most violent showers come from the huge cumulonimbus cloud; from below, these are dark and menacing but when viewed remotely from the side they often show a fibrous white anvil-shaped top. Lightning and thunder is a frequent feature of cumulonimbus clouds, as well as the heaviest and most severe forms of precipitation, like large hailstones.

Rain

'Rain' is produced by the mass uplift of air over a much larger area than convection currents cover. The cooling and condensing process is the same as for 'showers' above. The meteorologist will often refer to this sort of rain as dynamic precipitation, a general term that covers not only rain but drizzle, sleet or snow as well. This sort of precipitation generally falls from a 'stratiform' type of cloud, that is, much flatter and gentler cloud than the cumulus clouds from which showers fall. However, with mass uplift over a large area, there can be deep layers of stratiform cloud and heavy continuous rain may fall from leaden skies. In this situation, the atmosphere will tend to be in a 'stable' condition. 'Stable' means that the air is not spontaneously buoyant but is being forced to rise by other means.

There are two major mechanisms for lifting the air and producing dynamic precipitation:

- *Weather systems:* these are the features you see on weather charts, for instance 'lows' and 'weather fronts'. Both of these features generate rising air and cause unsettled weather conditions in their vicinity. Cloud and precipitation are very common.
- *High ground:* when an air flow is confronted by high ground it is forced to rise and cools. If the air is lifted above its condensa-

tion level, cloud will form. Then, precipitation may fall on the higher ground, especially on the windward (upslope) side. By contrast, the leeward (downslope) side is usually much drier because the air flows back down to lower ground and warms. In meteorology, the increase of rainfall on the windward side of high ground is known as 'orographic enhancement' while the dry conditions to the lee of high ground is referred to as the 'rain shadow'.

Do showers and rain occur together?

It is quite common to find regions of mass uplift with convective cells embedded within them. It all depends upon the structure of the atmosphere and on the meteorological developments that are taking place. There are many reasons why this should occur; almost all are linked to weather systems but these effects may be further enhanced by the high ground effect described above. Weather systems and high ground together can produce heavy and persistent precipitation.

Rainfall and flooding

There is a direct and obvious link between rainfall and flooding but extremely heavy rain will cause only limited and temporary flooding unless the water is contained. In simple terms, a large hollow in the ground surface will collect rainwater like a basin and retain it there. In addition, that particular basin will also collect water from other high ground areas as rainwater drains down channels from higher regions above. When this model is extended to a very large area which has a network of rivers or streams within it, the basin at the bottom will be filled very quickly.

What then happens to the collected water depends upon the geology of the basin. How permeable is it? Does it have run-off channels? Many ground surfaces are permeable and will absorb water but eventually the ground becomes saturated and no further absorption can take place. If the basin has one or more run-off channels, water rising above those levels will drain off to flow to another basin lower down. The lowest basin is usually the sea.

The way the water drains down will be determined by the quantity of water and by the shape of the channels. If there is a vast quantity of water and broad, flat river channels, flooding can extend over a wide area; such situations are seen routinely in countries affected by tropical monsoon conditions, with whole regions becoming deeply flooded. However much less rainwater can still cause flooding when steep and confined river channels collect lesser rainfalls and cause serious floods at downstream locations.

There is one other weather factor which is very relevant to flooding and that is the melting of accumulated snow. This is of course a winter phenomenon but it is significant that factors relating to ground permeability and shape do not apply to snow to the same extent as they do to rain. Heavy snowfall will accumulate on the top of most ground surfaces and even on ice sheets. The snow can then be banked into deep drifts by the wind. This means there is a large store of frozen water waiting to be turned into liquid at a time of thaw. If the thaw occurs coincidentally with heavy rain, the quantity of rainwater is greatly boosted by the snow-melt water. There are many occasions when melting snow has caused serious floods to occur, even without the addition of new water from rainfall sources.

Sea waves

Waves are water rotation movements in the sea around an approximately horizontal axis. These movements may be generated in any large body of water. The waves are the result of some wind energy being transferred into the water. The stronger the wind and the longer it blows over the water surface, the more energy is transferred and higher the waves become. Waves generated like this are known as 'wind waves'. In addition to wind waves there are large very flat waves known as 'swell (waves)'. Basically, these are the remains of wind waves generated elsewhere, often far away. Wind waves and swell combine together to give higher and potentially more dangerous waves.

When sea waves move into shallower water, the rotating patterns are disrupted and the effect of this is to build up each wave more steeply; this eventually causes it to 'break'. Sometimes sea waves can be built up to terrifying heights, causing coastal land areas to be flooded suddenly and catastrophically. Such waves can be produced to some degree by weather forces but more often are the result of disturbances on the sea floor, such as undersea volcanic eruptions or earthquake land heaves (See *Tsunami*). A major landslip into the sea can also cause huge waves. However generated, these huge waves are long-lasting and travel very quickly, suddenly and unexpectedly flooding coastal regions far from their point of origin.

Tides

The moon is a satellite of Earth; it rotates around it making one complete circuit every four weeks. It is held in its orbit by the balance of the Earth's gravitational attraction pulling it inwards and the centrifugal force of its curved orbit which acts outwards. These are exactly the same forces that hold the Earth in its orbit around the sun. The gravitation of the sun pulls the Earth towards the sun and the pull is balanced by the centrifugal force of the Earth's circular orbit. Effectively, the Earth and the moon form a fixed rotating unit around the sun and this means that the Earth has a slightly off-centre axis of spin.

The moon plays a very large part in producing the oceanic tides that affect seas and coasts everywhere in the world. The time between each high tide and the following high tide is around 12 hours 25 minutes, so a 'tidal day' approaches 25 hours. The generation of world tides lies within the combined effects of gravitation and rotation. The fluid nature of water means that it will respond to small changes in these forces. The major gravitational effect on the water comes from the Earth itself and that force holds the water on the planet's surface. But the seas (and everything else) are also affected by the gravitational effects of both sun and moon.

Because it is much closer than the sun, the moon's gravitational pull is the stronger and so its presence above any point on Earth opposes (reduces) the Earth's force of gravity at that point. The Earth's seas are therefore heaped up into a huge dome of water to create what is known as the 'direct tide'. The extra water required is supplied from those areas unaffected by the moon's gravity, that is, from the areas at the sides of the dome.

On the other side of the Earth, a similar dome of water (the indirect tide) is formed by the effect of the Earth's off-centre axis of spin. In this case, the mechanism involves the increase of centrifugal force. As with the direct tide, the extra water required is supplied from the sides. So the combined effect is the simultaneous generation of high tides on opposing sides of the Earth. As the Earth spins on its axis, making one complete revolution every 24 hours, the tidal effects appear to flow around the globe so all coastal regions receive a sequence of high and low tides.

A final complication is imposed by the addition of the sun's gravitational effects. When the sun comes in line with the moon, there is a stronger pull and the Earth's gravitation is reduced more than usual. This is the cause the higher tides called 'spring tides'. When sun and moon are in opposition, the tides are smaller and these are called 'neap tides'. The cycle of spring and neap tides is linked with one rotation of the moon around the Earth, that is, on a four week cycle.

Tornados

Tornados are intense low pressure features in which strong currents of air spiral into a central point near the Earth's surface and then spin violently upwards. They occur in association with conditions of deep instability where there are already extensive areas of unstable clouds. The air pressure becomes very low at the point of surface convergence and there is a considerable suction effect capable to lifting heavy items. The tornado assumes an inverted cone shape, often defined by the dust and debris within it or by water lifted from the sea. The dimension and power of the tornado depends on the degree of instability of the atmosphere and the wind flows which develop around it.

In very hot desert regions, tornados can form and exist in a dry form called dust whirls. They are caused by a very unstable but dry atmosphere and by very high surface temperatures. Although dust whirls can be powerful and extend upwards to several thousand feet, they lack the destructive force of tornados which are associated with cloud and moisture.

Severe damage can be caused by tornados, not only by the high winds in their circulations but by the powerful suction effect that forms at the point of surface convergence.

Tropical cyclones

The regional names for these huge and powerful weather systems are hurricanes, typhoons, severe tropical cyclones or severe cyclonic storms, depending on their location. All tropical cyclones form in regions of hot tropical waters, with over 85% of circulations forming near 20 DegN or 20 DegS. The tropical storm season of any region is their late summer months when sea temperatures are at their highest; a sea temperature of 27 DegC or over is usually required for formation to occur. Tropical cyclones are categorised according to their wind speeds; Category 1 is the weakest (75 mph) and Category 5 the strongest (over 155 mph).

Tropical cyclones form every year; worldwide, there are at least 80 per year. Their huge power is generated from the heat and plentiful moisture of the Tropics; as this very warm and moist air rises, it condenses and releases considerable 'latent heat'. This a very important mechanism in tropical cyclone development. Latent heat is the heat involved with change of state, that is, solid to liquid to gas and *vice versa*. In the atmosphere, water changes from ice to liquid water to water vapour and *vice versa.*In the hot atmosphere of the Tropics, when water vapour condenses to liquid water, considerable quantities of latent heat are released and upcurrents in the clouds are greatly boosted.

Although tropical cyclones are responsible for the most destructive weather on our planet, they are also an essential part of the global weather engine, helping to spread the heat of the Tropics to more northern latitudes.

Troposphere and Stratosphere

Planet Earth is encased by a deep layer of atmosphere which is held by the force of gravity. The atmosphere has a very distinctive temperature structure. Starting at the ground, the temperature decreases with height, losing several Centigrade degrees with every thousand feet. Then, at significant height, the temperature stops falling and stays relatively constant at even higher levels.

The lower part of the atmosphere where the temperatures decreases with height is called the Troposphere and this is the part of the atmosphere we all live in. When the temperature stops falling this is the start of the second layer of the atmosphere, called the Stratosphere. The level where troposphere finishes and stratosphere begins is called the Tropopause. The height of the tropopause varies considerably; it can be as low as 15,000ft or as high as 50,000ft.

This organisation happens because the structure of the troposphere is dominated by Earth/atmosphere interactions. Where these interactions stop extending upwards, this determines the level of the tropopause. Above this level, the stratosphere maintains its largely constant temperature because there are no Earth/atmosphere interactions taking place. Low tropopause levels are found where the troposphere is very cold – for instance near the Poles. A very high tropopause level is associated with a hot troposphere, characteristic of equatorial regions.

Tsunami

Tsunami is a Japanese word meaning 'harbour wave'. Tsunamis are huge sea waves generated by the energy released by undersea earthquakes. The most common and frequent locations for earthquakes are in the boundary regions of the Earth's 'tectonic plates'. Geologists have determined that the Earth's upper crust (the lithosphere) is divided into a number of rigid plates and that there is constant movement between them. These boundary zones of constant movement are called transform faults. Japan is close to one of the transform faults and therefore has experienced many tsunamis in the past; hence the adoption of a Japanese word for this feature into international scientific language.

The energy of an undersea earthquake causes the sea floor to rise and fall, lifting up the water above it and imparting the water volume with a huge amount of potential energy. This potential energy is translated into kinetic energy (the energy of movement) and tsunami waves spread out from the centre. Two types of tsunami wave are defined; the 'local' tsunami races off towards near coastlines while the 'deep ocean' tsunami makes off across the ocean. The speed of the tsunami wave is determined by the depth of the water so the deep water tsunami travels much faster. Tsunamis can travel at up to 800 kph (500 mph) in deep water and can still flood coastal areas after travelling many thousands of kilometres.

As the tsunami wave reaches shallower water, it heaps up to become much bigger; this is why there are many observations of a 'dark line' seen on the horizon. On its approach to the coastline, this heaped up wall of water does not break like a wind wave. Instead it arrives like a huge tidal flow. The strongest tsunamis may be 10m deep at this point. After flooding inland, the wave withdraws but further energy flows may cause it to return several times. Also, the earthquake itself is likely to produce aftershocks that may produce more tsunami activity.

Weather satellites

There are many natural satellites. Our world is a satellite of the sun. Our moon is a satellite of Earth. Some of the other planets in our solar system have satellites (moons) orbiting them. In all cases, the forces involved are the same. The satellite is attracted by the gravitational pull of the larger parent body but this force is balanced exactly by the centrifugal force which is generated by its curved path through Space. With the two forces balanced and virtually no friction to slow the satellite down, a stable orbit is achieved.

It was the middle of the 20th century before mankind was able to launch an artificial satellite into space. All earlier satellites were 'polar-orbiting' – that is, they circled the Earth in a north-south plane while the world rotated below them. Because these satellites were relatively close to the Earth they had to move at great speed so that their generated centrifugal force would be sufficient to balance the considerable gravitational pull of the Earth. A common height range was 600-800 km.

Weather satellites were first put in orbit in 1960. Today, there are a number of weather satellites that are polar-orbiting with highly sophisticated instruments on board; there are cameras to record visible imagery, infrared and water vapour distribution, sensory instruments to measure atmospheric profiles of temperature, humidity and wind, and special instruments to record other land and sea information, for example sea ice, wave height, vegetation distribution. These satellites make a complete orbit of the planet in less than two hours and the rotation of the Earth means they scan a different strip of our planet with each orbit

In addition to these polar-orbiting satellites, there have been for several decades 'geostationary satellites'. The idea of a geostationary satellite is to place it over the same spot so that it can measure changes continuously. This is achieved by locating the satellite in the position where it will rotate in unison with the Earth – the term used is 'quasi-stationary'. There is only one location where this can be done – over the Equator, since this is the only place where the forces involved could possibly be balanced. In order to make the satellite move in unison with the Earth, it has to be placed a considerable distance away – around 36,500 km. This means that its cameras and instruments can 'see' a complete half of the Earth. This is the source of the dramatic 'whole-globe' satellite imagery sometimes seen in illustrations or on television.

Obviously the accuracy of the measurements will decrease markedly towards the edge of the world images, because of the very oblique angles of the Earth's surface in such areas. This is solved by locating several geostationary satellites around the equatorial region, the optimum number being five. This is the current number of geostationary satellites in operation; two are operated by the USA, two by Europe and one by Japan.

Wind chill

By means of clothing (or lack of it!) we maintain the temperature around our bodies to the appropriate level to sustain our lives, continuously creating a little personal climate. The wind is one factor which determines how fast heat is removed from us; the temperature and the humidity level of the air are two other factors. If the wind is light, our 'climate' will be relatively undisturbed. If both the air temperature and the humidity are high, relatively little cooling will take place, as there will be little evaporation of our sweat. In these situations, we are likely to feel it very hot and muggy and the wind chill factor will be very low.

On the other hand, a strong, cold and dry wind will remove a great quantity of heat from us. The speed of the wind changes the air more quickly, the low temperature causes greater cooling and the low humidity means rapid evaporation of our sweat. The latent heat required for the sweat evaporation comes from our body heat and cools us even further. This would be a situation of significant wind chill.

It is possible to calculate what degree of wind chill may be associated with any wind flow and express it as a temperature change value. This is often given in winter weather forecasts, *viz:* 'The temperature will be 10 DegC but with wind chill it will feel more like 3 DegC.'

Wind

Wind is the movement of the gas that completely encases our planet. This gas, which we call air, is held in place by the force of gravity, a force of attraction that exists between all objects which have any mass (weight), however small they may be. Air is free to move in all dimensions and, in addition, is readily able to expand and contract.

There are two fundamental forces that determine all the complex movements of air. The most important of these is the heat from the sun. This stream of electromagnetic radiation heats up the surface of Earth and this heat is transferred to the air. The heated air becomes less dense and begins to rise; thus, currents (movements) are generated through the atmosphere.

Because our planet is a large globe, the sun's heat is much more concentrated in those surface areas which are presented squarely to the sun (the Tropics), while areas at increasingly high latitudes receive the heat much more diffusely (see Appendix B: Climate and Weather). This variation in heating generates huge convection-type airflows, with air rising from the Equator to great height, flowing towards the Poles, then sinking back to Earth in polar regiuons to flow back towards the Equator near the surface of the Earth.

However, this simple flow model is greatly complicated by another factor – the rotation of the Earth. The Earth rotates on its axis, making one complete revolution every 24 hours. This means that Equatorial regions are actually spinning at over 1,600 kph (1000 mph). The speed of rotation decreases towards the Poles but mid-latitude regions are still moving at considerable speed. The effect of this is to deflect the theoretical airflows so that their pole-ward movements are much slower and the wind flows organise themselves into vast undulating currents which move in all three dimensions.

Near the surface of the Earth, the wind is determined by the effect of these large-scale movements higher up but the exact wind speed and direction at any location and time will also be affected by a whole range of other factors, such as the nearby topography of the area, the type of land/water surface over which it is flowing and by the effects of human activity, too.